Value Management

The value management benchmark: Research results of an international benchmarking study

Steven Male, University of Leeds
John Kelly, Heriot-Watt University
Scott Fernie, Heriot-Watt University
Marcus Grönqvist, University of Leeds
Graeme Bowles, Heriot-Watt University

Thomas Telford
EPSRC
Heriot-Watt University Edinburgh
University of Leeds

Published by Thomas Telford Publishing, Thomas Telford Ltd, 1 Heron Quay, London E14 4JD. URL: http://www.t-telford.co.uk

Distributors for Thomas Telford books are
USA: ASCE Press, 1801 Alexander Bell Drive, Reston, VA 20191-4400, USA.
Japan: Maruzen Co. Ltd, Book Department, 3-10 Nihonbashi 2-chome, Chuo-ku, Tokyo 103
Australia: DA Books and Journals, 648 Whitehorse Road, Mitcham 3132, Victoria

First published 1998

A catalogue record for this book is available from the British Library

ISBN: 0 7277 2730 3

Typeset by Kneath Associates, Swansea
Printed and bound in Great Britain by Halstan & Co Ltd, Amersham, Bucks

Contents

List of figures

Executive Summary

This research report complements the document entitled *Value Management: The value management benchmark: a good practice framework for clients and practitioners*, the final output of the research project. Both documents are also available on an interactive CD-ROM (report and CD-ROM are available from Thomas Telford Publishing)

The research report is directed towards practitioners and academics currently involved in value management (VM). It is a reference document indicating the origins of the 'the value management benchmark', presenting the evidence and justifying the conclusions. The report therefore contains the research methodology, the corroboration techniques applied to the evidence collected, and the logic behind the conclusions drawn. However, it is not the intention of either document to dictate a prescribed VM process for blanket application throughout the construction industry, rather it is to synthesise current VM practice into a co-ordinated framework.

The need for the research report was first indicated at expert seminars held in Leeds and London, hosted by the research team and attended by VM experts from the UK. The purpose of the seminar was to critique and validate the VM benchmark document. Those attending the seminars indicated that a report detailing the research conducted would be a valuable document in its own right to allow the reasoning behind the research and its intended purpose to be brought out into the open. As the research was ongoing, various bodies in the UK produced a series of guidance notes. Those attending the seminars indicated clearly that the VM framework sat above these. These documents have been used as the primary comparative basis within the UK for setting the findings from the fieldwork in an international context.

The report begins by making explicit the context and environment when the research was proposed in 1995 and outlines the 'unknowns' at that time. The research methodology is then explained in full, highlighting the main aspects of the work. The research document is completed by identifying the issues addressed during the research, the future development of VM and areas for future research.

During the course of the research numerous texts, guidance notes and standards were consulted. The report uses references sparingly and readers wishing to review the material consulted can do so in a published bibliography of *Value Management* produced by the Building Research Establishment.

1. Introduction

The research project commenced with the recognition that the practice of value management (VM) was diverse nationally and internationally and across industry sectors. The research proposal that was submitted to the Engineering and Physical Sciences Research Council Innovative Manufacturing Initiative (EPSRC IMI) entitled *Construction as a Manufacturing Process* had four objectives. Three objectives have been met in their entirety. The fourth, concerned with contributing to a VM standard, became redundant because of parallel work undertaken in Europe. The resultant prime output *The value management benchmark: a good practice framework for clients and practitioners* compliments the European standard for VM which will be published by BSI in the near future. This research is the only substantive IMI grant awarded in the VM area that has established the only benchmark for VM recognised internationally.

The research work has drawn on the experiences and practice of 58 industrial collaborative partners internationally. It has also drawn on a diverse range of VM/value engineering /value assessment texts and related general management literature. In addition, Southern Water and Balfour Beatty were major sources of collaboration and assistance was given through the offices of the Society of American Value Engineers, the Institute of Value Management in the UK and the Institute of Value Management Australia.

The deliverables from the project are:

- *The value management benchmark: a good practice framework for clients and practitioners*, also available from Thomas Telford Publishing
- This detailed research report.
- A published bibliography, *Value Management,* produced by BRE.
- An interactive CD-ROM, *The value management benchmark*, also available from Thomas Telford Publishing.
- A WWW page on the project.

The research commenced in December 1995 and all fieldwork was completed by January 1998. The research team operated from two sites, one in the School of Civil Engineering at the University of Leeds and one in the Department of Building Engineering and Surveying at Heriot-Watt University.

The research report is broken down into three parts.

Part 1 sets out the context of the research and the methodology used in the study.

Part 2 sets out and debates the issues uncovered as part of the research study from fieldwork and through comparing the empirical evidence against a series of guidance notes and standards that were produced by various bodies while the research was ongoing.

Part 3 sets out the background to the benchmarking methodology used, its development and the final approach adopted.

PART 1: The Research Project Methodology

2. Context of the Research Project in 1995

2.1 Introduction

The research project commenced in December 1995. At that time the use of and interest in VM was gaining momentum throughout the UK construction industry, indeed, the Institute of Value Management (IVM) was undergoing a revival in its membership not experienced since the 1970s. However, there still did not exist a standard for VM, nor did there exist any formal training for VM practitioners. It was recognised that there was a wide diversity of VM practice in the UK construction Industry. In 1995 the following questions remained unanswered.

1.) How was the practice of VM conducted in various parts of the world?
2.) Could the VM process be mapped to allow comparisons across a wide diversity of VM practice in the UK construction industry and elsewhere?
3.) Could the VM process, particularly when adopted at the earliest stages of a project, be measured for its level of performance.
4.) Could the project adopt an 'off the shelf' benchmarking methodology, such as those described by Camp[1] and Boxwell[2].

2.2 Project Objectives

The original research proposal in 1995 had the following research objectives.
1.) To compare and contrast methodologies, tools and techniques of VM used internationally in construction and manufacturing against a benchmark methodology now implemented in the UK construction industry by the research applicants – Steven Male and John Kelly.
2.) To distil 'best practice' from benchmarking.
3.) To develop a best practice manual for VM in UK construction to assist in developing a standard.
4.) To test, validate and implement the best practice manual in the UK.

An important aim of the project at inception was to develop and implement a best practice manual for VM in construction, from pre-concept to the post-occupancy stages of a project.

1. Camp, R.C. *Benchmarking: The Search for Industry Best Practices that lead to Superiour Performance.* Quality Press, Milwaukee, 1989.

2. Boxwell, R.J. *Benchmarking for Competitive Advantage.* McGraw-Hill, New York, 1994.

3. The Adopted Research Methodology

The methodology adopted for the research (Figure 1) stayed very much within the remit of that submitted when the EPSRC Innovative Manufacturing Initiative sanctioned the project. It not only achieved the proposed objectives stated in section 2.2 but also exceeded them by including the publication of a research report to accompany the benchmark framework.

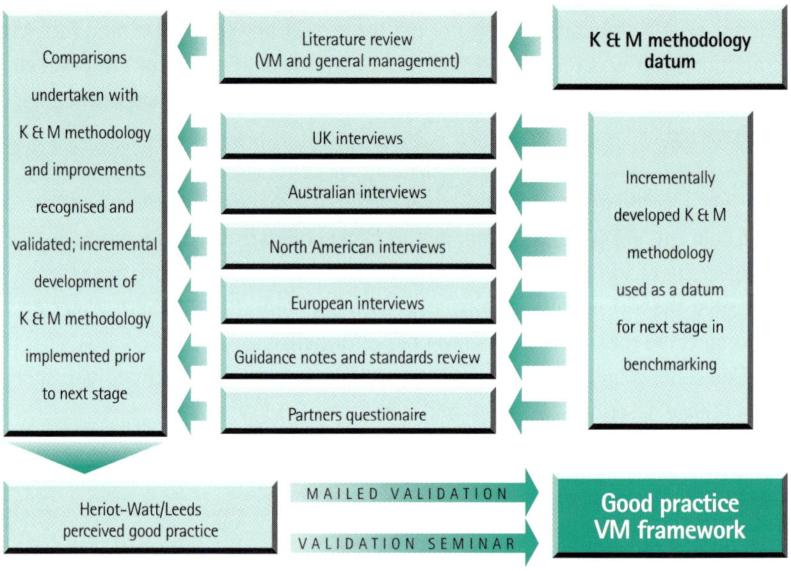

Figure 1 The adopted research methodology

The starting point for the research was the previous research work conducted by Kelly and Male in North America to give the team a base to work from and aid in the initiation of a literature review. This research work was documented as case study reports and tapes of interviews undertaken in North America.

The Kelly and Male (K&M) methodology, set out in more detail in Section 4, was adopted as an arbitrary benchmark datum to which the literature, fieldwork and other data were referenced in order to highlight disparate views and approaches to practice, procedures, methodologies and techniques used at stages throughout the project/product life cycle. Prior to using the datum, an analysis of VM workshop reports of projects facilitated by Kelly and Male was carried out to make explicit and map the VM process. This produced the detailed procedures, internal and external factors of influence, critical success factors, pre-requisites, tools and techniques used as set out in Figure 5 below. This would lend the datum to appraisal and comparison as per the research methodology.

Each stage in the methodology highlighted a number of issues for debate and these are detailed in the following sections of this report. Meetings were convened on a regular basis to discuss and analyse these issues and to incorporate any conclusions into the datum. This meant that the datum was continually updated throughout the research. The continuous updating required, on completion of all fieldwork, a very detailed questionnaire to be completed by all benchmarking partners as a final validation of research results. A framework document of the research team's perceived 'best practice' was compiled during the latter stages of the research programme. The first draft concluded the research team's perception of best practice in the current field of VM using the rich data collected from fieldwork.

The *Value Management Benchmark Framework* was tested against industry for validation purposes. This was achieved through collaborating partners and industry experts critiquing the framework document, either by mail and/or at expert seminars.

It is important to recognise that the framework document relates to construction. Considerable effort was expended in trying to undertake benchmarking with the manufacturing sector but with only isolated success. Manufacturing industry, taken as a whole, appears to regard VM as a management technique for competitive success and was generally reluctant to share details of specific tools and techniques. Construction VM was only found to be practised to any great degree in North America, Australia and the UK. Most of the successful benchmarking in construction was carried out in these geographic areas.

Notwithstanding this, a number of useful discussions were held with VM practitioners in the manufacturing sector in mainland Europe, UK, North America and Australia. Further an extensive review was undertaken of current texts and papers on the practice of VM in the Far East.

3.1 Fieldwork and Data Collection

Fieldwork was conducted in North America, Australia, UK and Europe using structured interviews and the benchmark 'datum' to elicit data on the practice of the VM process and to gain insight into a host of other relevant issues such as culture, legislation and certification. Fifty-eight structured benchmarking interviews were conducted during the course of the project; fourteen in Australia, eighteen in North America, twenty-one in the UK and five in Europe. The interviews revealed, as expected, a diverse range of current VM practice. Interviews were followed up with a summary of benchmarks established and latterly the substantive questionnaire to fill gaps in data revealed by the last series of interviews with benchmarking partners, as indicated above.

In the UK, best practice practitioners were identified by asking those interviewed which organisations they considered to be best in class. Of those interviewed the majority consistently cited particular organisations which were then targeted for data collection, if not already done. Due to the logistics involved in adopting a similar method for gaining access to best-in-class VM practitioners in both North America and Australia, an alternative method was adopted for these continents. In North America the research team attended the Society of American Value Engineers (SAVE) Conference in Seattle, opening the conference with a workshop designed to explain the research project and to recruit best-in-class practitioners as collaborators for the project. A similar workshop/seminar was held at the Institute of Value Management Australia (IVMA) for the same purpose.

4. The Kelly and Male VM Methodology

As a benchmarking research project it was necessary to set out a reference point or datum around which the benchmarking studies could be conducted. Unlike many benchmarking exercises, the datum in this instance is a VM methodology, developed through research studies conducted by Kelly and Male from the mid 1980s onwards. This has included its implementation, as practised by them, through live action-research oriented consultancy commissions conducted from the early 1990s onwards, commissioned by private sector organisations and government clients.

This section outlines the research datum used to conduct the benchmarking studies. With the experience of applying it in many diverse projects, the methodology has evolved into a rich framework of tools and techniques. Not all the techniques presented are used on every project but, in practice, those that are most appropriate for the project in hand are selected for 'live' projects. The section is broken down into two main sub-sections, the methodology as developed up until the benchmarking study of 1995 and that developed as part of the research work.

4.1 Research Leading to the Benchmark Text in 1993

The methodology developed prior to the benchmarking study was derived from a number of research studies and has been set out in a series of reports, papers and books.

4.1.1 1986 - The First Visit to the USA

This first visit to the USA was funded by the Education Trust of the Royal Institution of Chartered Surveyors (RICS) and followed from a number of literature review studies on the application of North American value engineering to UK construction. The primary research question to be answered was "is value engineering another name for the cost planning/control service offered by the UK quantity surveyor?" The research undertaken by Kelly and Male took in a wide spectrum of practice in USA and Canada and concluded that the service of value engineering was different to quantity surveying practice in a number of significant areas. It was also noted that practice on the ground was considerably more varied than that portrayed in the value engineering texts. A detailed report was submitted to the RICS and the following publications produced.

- Kelly J.R. and Male S.P. *A Study of Value Management and Quantity Surveying Practice*. Surveyors Publications Ltd., 1988, published under Occasional Papers cover, ISBN 0 85406 398 6.
- Male S.P. and Kelly J.R. Organisational responses of public sector clients in Canada to the implementation of value management: lessons for the UK construction industry. *Construction Management and Economics*, 1989, **7**, 203-216.

4.1.2 1989 - The Second Visit to the USA

The second visit, conducted in 1989, was also funded by the Education Trust of the RICS . The 1986 research left a number of questions unanswered: first, the techniques used during early design or orientation; second, specific techniques used for function diagramming; and third, techniques used during a 40 hour VM workshop. Further funding was obtained to make a return visit to the USA. During this project, data were gathered in all three research areas mentioned earlier and the following publications produced.

- Kelly J.R. and Male S.P. *The Practice of Value Management: Enhancing Value or Cutting Cost?* RICS, 1991, ISBN 0 85406 493 1.
- Male S.P. and Kelly J.R. The economic management of construction projects - an evolving methodology. *Habitat International*, 1991, **14**(2/3), 73-81.

4.1.3 1991 to 1992 - A Two-Part Study of Client Briefing

Research results from the 1989 study had left questions relating to the application of VM at the early design or orientation stage unanswered. The subsequent study was again funded by the Education Trust of the RICS and led to a study in two parts. The first was to review the methods of briefing and the second to propose an alternative approach based upon the techniques of VM. The primary outcome from this research was the recognition of a two-stage briefing process, one dealing with the 'business brief" for a project and the second dealing with the 'technical brief' for the project. A practice manual was produced for the briefing process identified below, which explored in more detail how to conduct a two-stage briefing process using VM. The proposed techniques were tested in an expert seminar hosted by the RICS and attended by representatives of influential client organisations. Three publications resulted.

- Kelly J.R., MacPherson S. and Male S.P. *The Briefing Process; A Review and Critique.* RICS, 1992, ISBN 0 85406 541 5.
- Kelly J.R., Male S.P. and MacPherson S. *Value Management - A Proposed Practice Manual for the Briefing Process.* RICS, 1993, ISBN 0-85406-631-4.
- Kelly J.R., MacPherson S. and Male S.P. Functional levels in building design. *Proceedings of CIB W55 Symposium - Economic Evaluation and the Built Environment*, Lisbon, Sept 1993, **4**, 115-125.

4.1.4 Publication of the Text to Become the Forerunner of the Benchmark

By the end of 1992 sufficient material had been generated by research for the publication of a book, which was to become the forerunner for the benchmark for this study. The book took the theories and principles of the North American work and argued the case for a structured approach to VM in a UK construction context. The resulting text was:

- Kelly J.R. and Male S.P. *Value Management in Design and Construction.* E & FN Spon, 1993, ISBN 0 419 15120 6, 0 442 31611 9 (USA), 181pp.

4.1.5 The 1994 Study – VM and Strategic Project Decision Taking

Further funding was obtained from the RICS to investigate the application of VM to strategic decision taking on projects. This study was prompted by questions remaining from the briefing study. Contacts in North America, having read the RICS reports, commented that the practice of 'architectural programming' was significantly similar to the approach being proposed in the 1993 publication for this to be worthy of investigation. A further investigation raised a number of issues that have been addressed in the current benchmarking study. The 1994 study was recorded in a paper given at the first RICS Construction Research Conference, COBRA 1995, and entitled:

- Kelly J.R. and Male S.P. Facilities Programming. *Proceedings of COBRA'95 RICS Construction and Building Research Conference*, Edinburgh, Sept 1995, **1**, 99-106.

4.2 The Pre Benchmarking Study Methodology up to 1995

This section summarises the Kelly and Male methodology as developed up to the period before the award of the benchmarking study by the EPSRC IMI programme.

Kelly and Male[3] identified four levels where a workshop study may fit in to the project delivery process. These are detailed below in Figure 2. It was presumed that these levels would have a direct impact on how a VM workshop is carried out. The levels are:

- Level 1: task
- Level 2: spaces
- Level 3: elements
- Level 4: components.

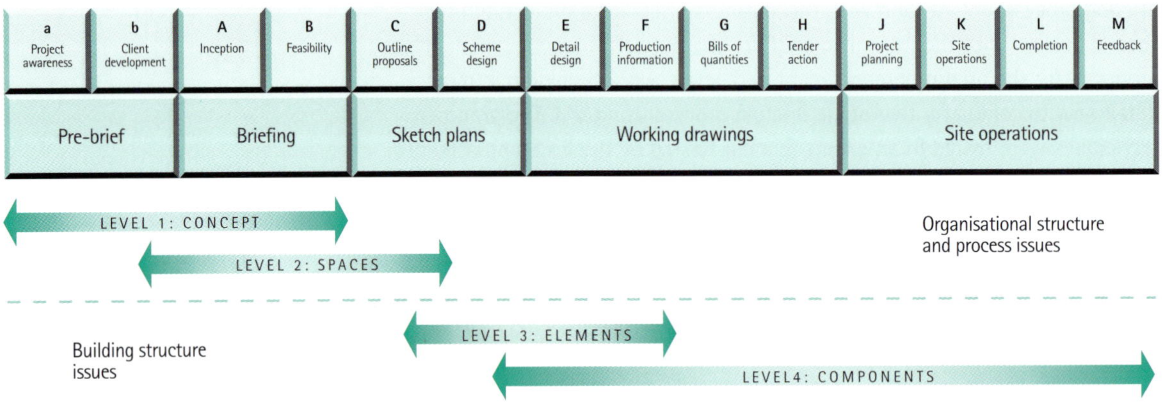

Figure 2 Decision levels and the project life cycle

Nine different VM workshop study opportunities were identified (excluding normal quantity surveying services that could be chosen as part of a project economics service). The choice of approach depended on the client problem to be addressed and the stage of the project life cycle that was to be targeted for study. Once a study approach has been appropriately selected, certain tools and techniques will be applicable to that study. Five basic approaches to undertaking a VM study were identified and are illustrated in Figure 3. These were termed:

1.) *VM1*, where the personnel involved are a study facilitator and client representatives who have an input to the definition of the project. The primary task of this team is to identify and state explicitly in functional terms: the project task, client needs and wants. The team will primarily be concerned with Level 1 but may address Level 2 issues. The aim of this study is to ensure that the client requires a built solution.
2.) *VM2*, where the personnel are a study facilitator and client representatives together with an independent audit design team. This design team may not be involved with the project design. At this point the decision to build has been taken but the procurement route is still open. The primary task of this team is to address Level 1 and Level 2 issues. This allows the procurement route to be the subject of investigation when strategic project issues are explored with the client. It may also lead to faster procurement since the project may be appraised functionally and technically prior to the brief being detailed for design.

3. Kelly J.R. and Male S.P. *Value Management in Design and Construction.* E & FN Spon, 1993.

3.) *VM3*, where the personnel are the study facilitator and client representatives together with the project design team. At this point the decision to build has been taken and the appointment of a project team would indicate a traditional procurement route but with the options for construction management or management contracting still open. This is seen to be non-conflict oriented and with increased possibilities for implementation. The exercise can be used for briefing the design team and maximises project knowledge. This exercise is seen to be inexpensive since the design team is not likely to charge an additional fee for this briefing exercise.

4.) *VM4*, where the personnel involved are a study facilitator together with the project design team. The project will have reached sketch design. While the team may address Level 2 issues, in the absence of a client representative they are more likely to be concerned with Levels 3 and 4. The absence of the client would indicate that the client's value system is embedded in the design to date. This is non-conflict oriented but can increase the design period, particularly where reference back to the client is required.

5.) *VM5*, where the personnel involved are a study facilitator together with an independent audit design team. The project will have reached sketch design. While the team may address Level 2 issues, in the absence of a client representative they are more likely to be concerned with Levels 3 and 4. The absence of the client would indicate that the client's value system is embedded in the design to date. This provides objectivity and the opportunity for increased technology transfer. However, there is an increased risk of adversarial relationship developing and can also increase the design period.

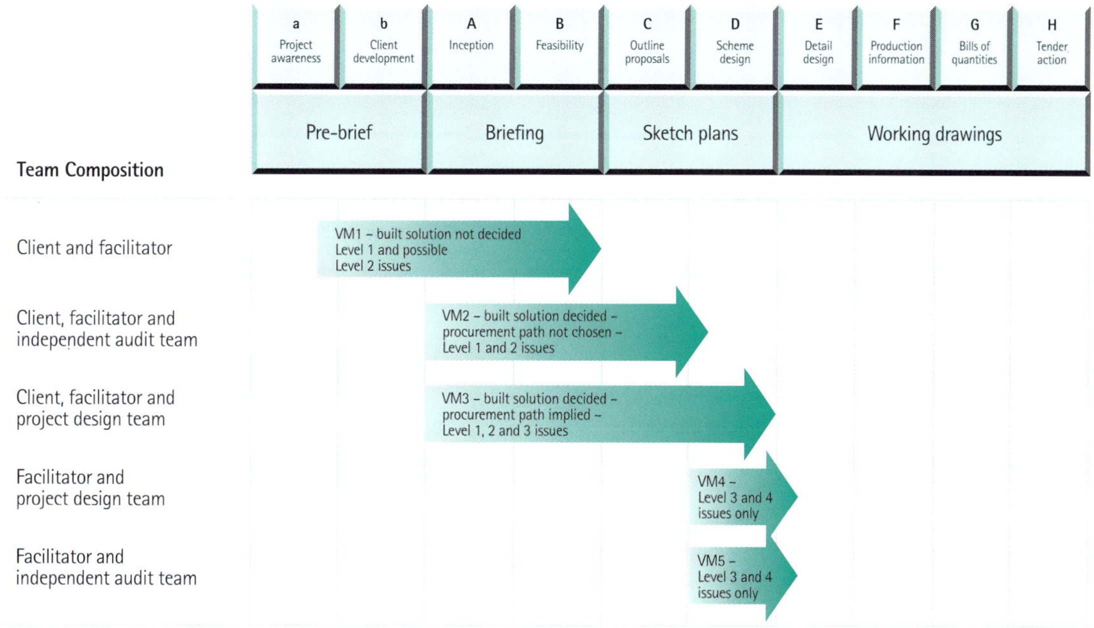

Figure 3 The five basic approaches to VM

Four basic approaches to undertaking a structured cost study were identified and are illustrated in Figure 4. These are termed:

1.) *SCS1*, where the personnel are a study facilitator and client representatives together with an independent audit design team. This design team will not be involved with the project design. The project will have reached sketch design stage. This provides objectivity and the opportunity for technology transfer. However, adversarial relationships with the existing design team are a possibility. The primary task of this team is to address Level 3 and 4 issues. However, there may be small changes to Level 2 issues.

2.) *SCS2*, where the personnel are a study facilitator and client representatives together with the project design team. The project will have reached sketch design. This is non-conflict oriented but the incentive for change is limited. The primary task of this team is to address Level 3 and Level 4 issues. However, there may be small changes to Level 2 issues.

3.) *SCS3*, where the personnel involved are a study facilitator together with the project design team. The project will have reached sketch design and the team will address Level 3 and 4 issues. The absence of the client would indicate that the client's value system is embedded in the design to date. This is non-conflict oriented but the incentive for change is limited.

4.) *SCS4*, where the personnel involved are a study facilitator together with an independent audit design team. The project will have reached sketch design and while the team may address Level 2 issues, in the absence of a client representative are more likely to be concerned with Levels 3 and 4. The absence of the client would indicate that the client's value system is embedded in the design to date. This is potentially adversarial but is objective and with the opportunity for technology transfer. However, the implementation rate of potential savings tend to be low and there is a danger that quality can be compromised. Equally, a quantity surveying cost study is most likely to address Level 3 and 4 issues as illustrated following the completion of the sketch design and the cost plan.

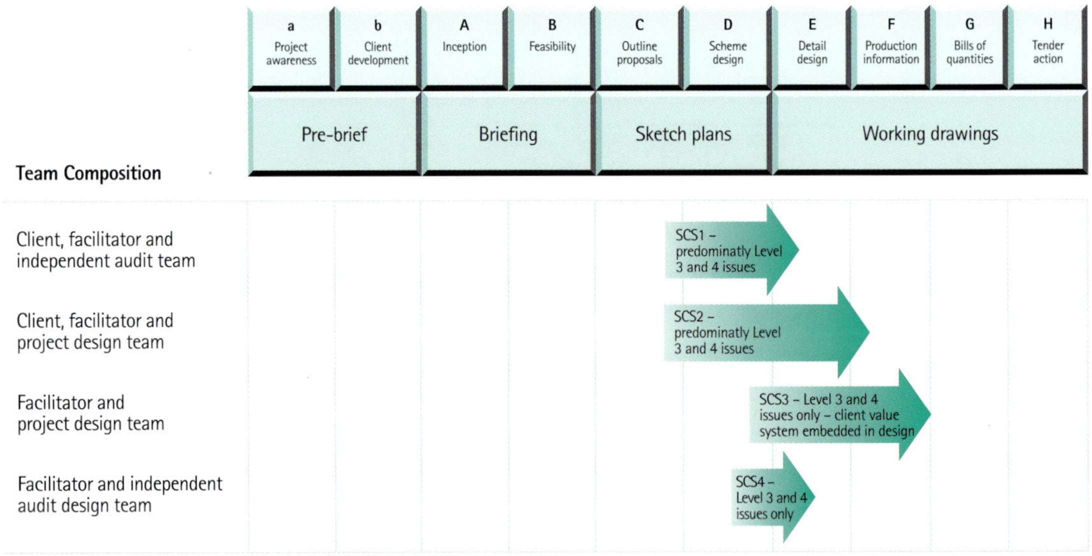

Figure 4 The four basic approaches to cost management

From the previous texts cited above, the methodology up to this point in time can be summarised as follows.

- Associating VM with the management of projects and the link between the latter and corporate and business strategy. VM is, therefore, a business term that describes a philosophy, service and approach that ensures the 'value thread' is maintained throughout the project life cycle. It achieves this through VM or cost study workshops, as points of convergence in the

project where there is a need to bring different skills and knowledge together to ensure value-for-money.
- Identifying the possibility of both VM and cost management studies, with VM studies occurring earlier in the project life cycle.
- Defining points for value opportunities in the project in terms of different levels for a study and identifying the characteristics of each study.
- Identifying some of the inputs and outputs for each of the VM and cost management opportunities.
- Identifying two stages to the briefing process, resulting in a policy brief and a concept brief as the outputs.

4.3 The 1995 EPSRC IMI International Benchmarking Study

To establish a datum for the research project, Kelly and Male made explicit their methodology by a detailed walk-through of the processes, tools and techniques they adopt on various types of VM study. This has been consolidated schematically into the datum for the benchmarking exercise and is set out in Figure 5.

This was supported by an in-depth review of live VM exercises that they had conducted. The analysis of a series of live project case studies facilitated by Kelly and Male revealed that the actual tools and techniques applied in a workshop depend on a number of factors. These include:
- the nature of the client and motives for carrying out the exercise
- the type of project and its stage in the design process
- the composition of the VM team
- the practical constraints imposed on the workshop itself.

From this review the process of VM could be broken down into three main areas, the *inputs* required for a workshop, the *process* of VM during the workshop, and the *outputs* of VM in terms of what the client is left with when the workshop is over. At each stage of the workshop a number of steps are undertaken to progress the job-plan satisfactorily. For each step there are a number of techniques available, the most appropriate depending upon the circumstances.

Figure 5 shows these steps in the shaded boxes under the various job-plan stages, with a selection of techniques that have been employed in practice. It is not the intention here to describe the classic job-plan stages. The schematic was used as the basis of the detailed dialogues with benchmarking collaborators on the research project.

4.4 Some Remarks on the Methodology

The development of the VM methodology has occurred in three distinct phases. Phase 1 stemmed essentially from a number of research projects investigating North American value engineering practice and postulating how this could be adapted into a UK context. This work culminated in the publication of *Value Management: the economic management of projects*, published by E & FN Spon. Phase 2 involved conducting concurrently sponsored research projects and commissioned action-research consultancies that enriched the methodology through the integration of theory and practice. Phase 3 involved making explicit the methodology as the benchmarking datum for the EPSRC IMI project. This was the first time that the VM methodology had been made fully explicit.

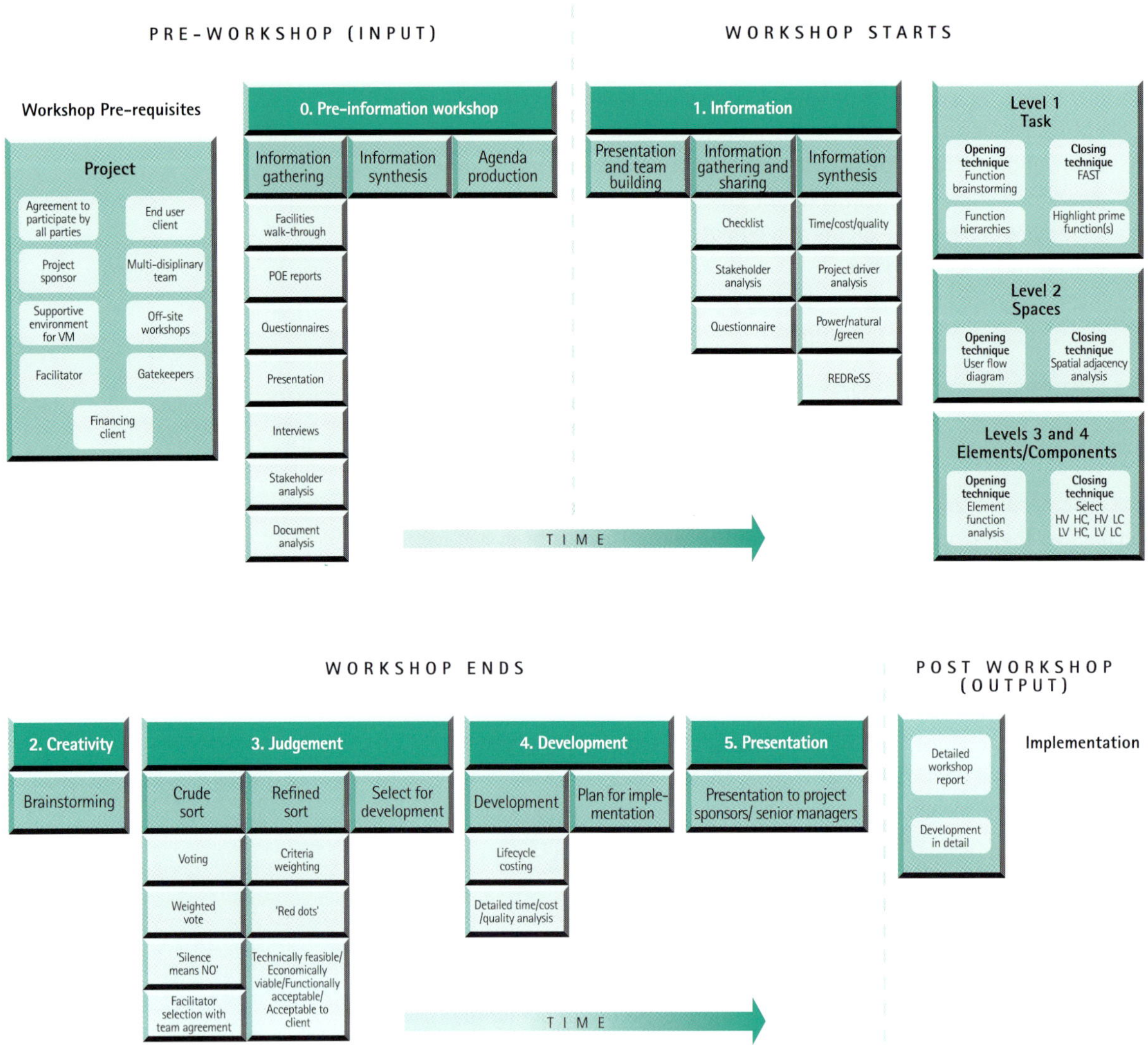

Figure 5 The Kelly and Male Value Management Methodology

The elicitation of the methodology into an explicit framework indicated a number of issues for further consideration.

- The information stage is often cited as the most important stage in the VM process, laying the foundations on which the remainder of the exercise is built. The process diagram shows that the information stage has the greatest number of possible activities associated with it, particularly when pre-workshop information is taken into account.
- It is notable that there is no variation of techniques applied at the creativity stage. Although the literature reveals there are a fair number of creativity techniques,[4] brainstorming proves to be by far the most popular in practice.
- Although Figure 5 shows the process as a sequential series of well defined steps, the practice of VM is not so clear cut. There is overlapping and iterating of the team's activities.
- The exploration of the Kelly and Male methodology also highlighted that fact that the concept of levels is robust regardless the type of project - whether civil engineering, building, process plant, etc. However, civil engineering and process plant studies were similar in that workshops were normally addressing Levels 1, 2 and/or 3 whereas workshops on building projects encompassed Levels 1 to 4.

4. Parker, D. E. *Value Engineering Theory.* McGraw Hill, New York, 1985.

PART 2: Issues Addressed by the Research Project

5. Issues Addressed by the Research Project

5.1 Introduction – A Review of Literature, Guidance Notes and Standards

In light of making the Kelly and Male VM methodology explicit, it was necessary to:

- identify published literature from the academic and practitioner community[5] and publications from government and professional bodies in the from of guidance notes and standards on the application of VM. These were being published as the research proceeded
- critique these documents in an attempt to highlight any differences in the approach, methodology and application.

The guidance notes and government standards that were consulted are highlighted below.

Guidance notes

- Building Research Establishment. *Value from Construction - Getting Started in Value Management.* BRE.
- Institution of Civil Engineers. *Creating Value in Engineering.* Design and practice guide. Thomas Telford, 1996.
- Construction Industry Research and Information Association. *A Clients' Guide to Value Management in Construction.* CIRIA
- Building Services Research and Information Association. *Value Engineering of Building Services.* Application Guide. BSRIA.
- HM Treasury Central Unit on Procurement. *No. 54 Value Management.* HM Treasury.
- Construction Industry Board. *Fact sheet on - Value Management.* CIB 1997

Government standards

- European Committee for Standardisation. *European Standard - Value Management.* CEN
- Standards Australia and Standards New Zealand. *Australian/New Zealand Standard - Value Management.* AS/NZS.

Other standards

- SAVE International 'The Value Society'. *Value Methodology Standard*
- National Centre for Value Management. *Competency Standards for the Value Management Facilitator.* University of Canberra.

The academic and practitioner literature was consulted and the above notes and standards have been compared, contrasted and synthesised to draw together VM process issues, intervention points, tools, techniques and procedures for incorporation into the benchmark framework. As a consequence, other issues were also highlighted from the fieldwork for further exploration and debate.

5.2 The Global Development of VM

5.2.1 The Origins of VM

Value analysis was developed by Lawrence Miles as an organised approach to providing the necessary functions at the lowest cost. It began in the USA

5. The literature consulted as part of the research project has been published by the Building Research Establishment as a bibliography.

manufacturing industry of the 1940s and remained embedded in manufacturing through the next two decades.[6] The practice in manufacturing remained fairly true to Miles' original methodology of:

- analyse the component parts of a product in terms of function
- consider ways of providing the functions at a lower cost
- confirm the economic and technical viability
- change production procedures.

Its global spread was mainly through multinational manufacturing companies and through the activities of individual American practitioners of value analysis. The idea spread to the UK in the early 1960s leading to the establishment of the IVM. A similar spread within manufacturing took place in, for example, Japan, Australia, France and Germany. The research study uncovered some of the diffusion process of VM for certain countries and this is set out in more detail below in Section 5.2.3.

5.2.2 Practice Versus Theory in Construction

Fieldwork in the USA, Australia and the UK indicates that the actual practice of VM in the market place differs from the theory depicted in texts and institutional frameworks for certification. This is particularly marked in the USA, where theory and practice in government studies favour the five-day independent team approach. However, in the USA private sector there is a diversity of practice, possibly due to the competitive nature of the VM market place. VM practice in private sector construction in that country is not dissimilar to that found in the UK or Australia.

5.2.3 The Construction Versus Manufacturing Debate

The benchmarking study revealed that VM/VE/VA in manufacturing has gone 'underground'. It is seen as a method that can provide a competitive advantage. It forms part of a broader 'tool-kit' of management techniques that are applied by manufacturing companies to provide a competitive edge and enhance value-for-money for the customer. In manufacturing, VM is mainly components-oriented although some representatives from manufacturers interviewed as part of the study wanted to bring VM forward into the early stages of product development. VM in construction, on the other hand, is a separate service, which is used at various stages throughout the project life cycle.

It can be argued that there are similarities between the management of the in-house VM programmes adopted by manufacturers and those for large organisations which are experienced clients of construction, who have implemented it as part of their formal project delivery process. A contrasting difference between regular users of VM in manufacturing versus those in construction is that the former will often spread a study over a number of weeks or months, whereas in construction the tendency is for intensive single interventions on one project.

The following subsections set out the process of diffusion from manufacturing to construction in the USA, Australia and the UK.

From manufacturing to construction: USA experience

The involvement of the US military in 1954 marked the point at which value analysis moved out of the manufacturing sector. Miles' original term of 'value analysis' was changed at this time by the US military to 'value engineering' for

6. Miles L.D. *The Techniques of Value Analysis and Engineering.* Lawrence D. Miles Foundation, 1972, 3rd edn.

the administrative reason that engineers were considered the personnel most appropriate for this programme. Table 1 demonstrates the growth of value engineering within US government agencies.

The focus of value engineering in construction in the USA is in the public sector and the Society of American Value Engineers, renamed SAVE International, has lobbied continuously for legislation requiring value engineering to become a mandated part of public sector project development.

Table 1: The growth of value engineering in the USA public sector.

Year	Development
1954	Introduced into US Department of Defense, Bureau of Ships
1963	Introduced into US Department of Defense, Navy Facilities Engineering Command
1965	Introduced into US Department of Defense, Army Corps of Engineers
1968	Used by Facilities Division of the National Aeronautics and Space Administration
1973	US General Services Administration, Public Building Service published the first value engineering service contract clauses for use in design and construction management contracts
1974	US General Accounting Office publish (May 6, B-163762) *Need for increased use of value engineering, a proven cost saving technique, in federal construction*

In June 1993 the US Office of Management and Budget issued a circular calling for "government wide use of Value Engineering and requiring Federal agencies to implement Value Engineering techniques". On 10th February 1996, President Clinton signed the Defense Authorisation Act, now known as Public Law 104-106, which contains the following:

> Sec.36 VALUE ENGINEERING
> (a) IN GENERAL. - Each executive agency shall establish and maintain cost-effective value engineering procedures and processes.
> (b) DEFINITION. - As used in this section, the term value engineering means an analysis of the functions of a program, project, product, item of equipment, building, facility, service, or supply of an executive agency or contractor personnel, directed at improving performance, reliability, quality, safety, and life cycle costs.

The US General Accounting Office estimates that agencies can save between 3% and 5% of a programme cost through value engineering activity. Amounts of savings by Federal departments in the financial year 1995 are given in Table 2 (SAVE International, 1998).

Although full legislation is in place, the take up by all government agencies is slow. A programme with a high implementation rate is that run by New York City Office of Management and Budget (OMB), the Treasurer's department.[7] In this situation a VM team within OMB receives details of all bids for capital works from the various city agencies. An assessment of the VM activity is made and contracts let to private sector certified (SAVE International) value specialists. The project is then tracked and VM studies undertaken. It should be noted that the procurement and supply of VM services is by OMB.

7. Kelly J.R. and Male S.P. *Value Management in Design and Construction: the economic management of projects.* E & F.N. Spon, 1993.

Table 2 Estimated savings to be attained through value engineering

Federal department	Potential saving: $
Defense Department	734,385,000
Department of Transportation	686,373,874
General Services Administration	109,608,453
Army Corps of Engineers	59,554,000
Department of the Interior	22,427,840
Department of Agriculture	8,764,155
Justice Department	5,990,387
Veterans Affairs	2,270,800
Health and Human Services	1,884,464
Agency for International Development	800,000
State Department	91,721

Value engineering in United States manufacturing has a strong following but value engineering teams are predominantly in-house and their activities subject to commercial confidentiality. The value engineering programmes of manufacturers, interviewed during the 1997 study, were a key feature in their strategy to gain a competitive edge. While the basic philosophy of Miles was being used to effect, value engineering had incorporated techniques more aligned to strategic planning, for example quality function definition, functional specification, technology road-mapping, etc.[8]

From manufacturing to construction: the Australian experience

Value engineering came to Australia in the middle to late 1960s through the activities of multinational companies such as Hawker Siddley and the marketing activities of US value engineering practices such as Value Analysis Inc.[9] In Australia it was recognised that the activity was more management than engineering in orientation and the IVMA became the value society. VM in Australian manufacturing and in the defence sector has maintained activity but the IVMA went into decline such that by the mid 1980s there was little activity.

In the mid 1980s, Roy Barton, a construction academic at the University of Canberra, undertook a study tour of value engineering practitioners in the USA. He returned to Australia with the message that while some changes to the US system were necessary to reflect cultural differences, the basic service was applicable to Australian construction. At the same time the US practice of Smith Hinchman and Grylls was undertaking value engineering for Leighton Construction. This led to the development of two strands of the same VM message.

In parallel with these activities, the New South Wales Government was determining a strategy for reducing borrowing and securing across-the-board gains in productivity. In this respect in was recognised that a 10% improvement in construction efficiency would improve the gross domestic product by 2.5%. In 1992 it was the government's wish that the industry's productivity in New South Wales be lifted by 20% within five years.

The strategy in respect of construction is embodied in the *Total Asset Management System Manual* (NSW Government, 1993), which incorporates a 28-page section on VM. The following extract details the procedures:

8. Shillito M.L. and Marle D.J. *Value: Its Measurement, Design and Management.* John Wiley and Sons. New York.

9. Adams E. *Value Management: cost reduction strategies for the 1990s.* Longman, 1993.

> In undertaking its project improvement role, the Ministerial Capital Works Committee will ensure that Value Management has been applied. To this end, reporting requirements that parallel those for Economic Appraisal, have been adopted for Value Management.
>
> These procedures are as follows:
>
> ***Projects less than $5million***: No formal requirements exist in terms of project submissions to the Committee. However, Value Management techniques should be applied, particularly in establishing the project rationale and considering options.
>
> ***Projects of $5million and over***: Formal Value Management Studies are required and submissions to the Committee require a summary of the Value Management Study outcomes, copies of the Value Management Study reports, and the agency's preferred direction and implementation strategy.
>
> ***Designated projects***: Due to the importance and/or sensitivity of certain projects, the Committee may attach specific Value Management assessment and reporting requirements.

The manual also states that the Ministerial Capital Works Committee will maintain a panel of private sector VM specialists for the New South Wales public sector agencies. Members of the panel are required to have successfully undertaken a facilitation course offered by the University of Canberra. Although the IVMA has four levels of membership, it is not known whether these are aligned to a level of qualification or competence.

During interviews with VM practitioners in Australia it became apparent that the various agencies interpreted systems outlined in the manual in a different manner. Some practitioners admitted that the system in some agencies was no more than a half-day facilitated meeting so that the VM box could be 'ticked'.

From manufacturing to construction: the UK experience
Value engineering came to the UK in the early to middle 1960s, it is presumed through the activities of multinational companies. Similarly to Australian events, it was recognised that the activity was more management than engineering in orientation and the IVM was formed in 1966. VM in manufacturing went largely underground as those companies which maintained activity drew their VM teams within their organisation and concentrated upon identifying a competitive edge. The IVM went into almost terminal decline such that by 1989 there were fewer than ten members. During the latter half of that decade the European Commission's DG XIII initiated a project under the Strategic Programme for Innovation and Technology Transfer (SPRINT). The aim of the project was to compile the basic techniques and methodologies required for a European standard on VM. The IVM became a partner within the SPRINT programme and as interest in the concept grew the IVM was rejuvenated.

In 1986 Kelly and Male undertook a research study funded by the RICS Education Trust entitled *A Study of Value Engineering and Quantity Surveying Practice*. A 133-page report under the same title was submitted to the RICS and an occasional paper was subsequently published.[10] The main conclusions of the report are given below.

- The etymological root of the term 'value engineering' lies in North

10. Kelly J.R. and Male S.P. *A Study of Value Management and Quantity Surveying Practice.* Occasional Paper. Surveyors Publications Ltd, 1988.

American manufacturing. 'Value management' is a more acceptable term for UK construction.

- There is a place for value engineering in the UK construction industry. The form it will take should be adapted to suit UK practice.
- In practical terms, if a decision is taken now to add value engineering to the surveyor's portfolio of services it is unlikely to have a major impact on the UK construction industry in the next five to ten years.
- Any further discussion on value engineering should concentrate not on whether it should be introduced but in what form, by whom and when it should be institutionalised.
- Value engineering is team orientated and therefore the value engineering team co-ordinator is not expected to be an expert in all technical areas.
- Value engineering provides a method of integration in the building process that no other management structure in construction can provide.
- Value engineering complements project management.

As predicted, VM in construction was slow to take root. Government agencies in the main did not take on board the ideas nor recognise any benefits. One London quantity surveying practice was offering a value engineering service during the 1980s to a limited number of clients. Certain major construction companies were also bringing value engineering into the UK through their links with North American construction organisations. Expansion in construction came from the newly privatised organisations, such as BNFL, BAA, Railtrack, BA and the water companies, particularly Yorkshire Water and Southern Water, and now activity is more widespread although at a low level. London Underground, in the public sector transport industry, has been actively promoting the use of VM.

A number of construction organisations have published guidance documents, for example BRE, CIRIA, BSRIA, ICE and CIB. HM Treasury's Central Unit for Procurement published its note Number 54, primarily applicable to the construction industry. However, there has been no compulsion by government to require agencies and departments to use VM as there has been in the USA and Australia.

5.2.4 Procedures Adopted in Other Countries

A number of countries, for example Japan, India and South Korea, adopted the SAVE International model of training and certification and have reciprocal arrangements. India has a national standard, *IS 1180-1986: Value Analysis - Glossary - Work Plans.* The existence of a standard in Japan and South Korea is unknown. In Europe, value societies were formed in Germany (1974) and in France (1978) and both of these countries have national standards for value analysis/management. Although some construction VM is undertaken, the primary focus is on manufacturing. These two societies are active, as witnessed by their joint production of the *Value Management Handbook* published by European Commission's DG XIII (EC, 1995).

5.2.5 Conclusions

The origins of VM are rooted in the United States manufacturing industry of the Second World War era. Diffusion into the manufacturing industries of different countries appears to have occurred during the 1960s and 1970s, often with the establishment of VM societies, in one form or another. Some of these societies are offshoots of the Society of American Value Engineers. There appears to have been a decline in the activity of some of these societies since this period, apparently due to VM in manufacturing going 'underground' as part of the competitive armoury of management techniques that manufacturing organisations use to combat global competition. This has been confirmed through fieldwork undertaken during this project.

VM is developing differently in Europe than in other countries and has been given a major boost by the European Union through the establishment of the SPRINT programme. VM has gone through a revival with its use in construction, often giving VM societies a new lease of life and with construction personnel dominating the membership. The Society of American Value Engineers, now renamed SAVE International, has lobbied effectively to make VM a formal part of government activity and this has now been enshrined in law. This will almost certainly have an effect on the continued spread of VM in the USA and in the requirements for certification and professional development. Many countries are also establishing VM standards, in particular Europe is addressing this at EU level.

VM has diffused into academia to different degrees in certain countries and this may have an impact of the relationship between VM societies and their respective educational sectors. In manufacturing, the diffusion of VM to the UK and Australia is summarised in Figure 6 below.

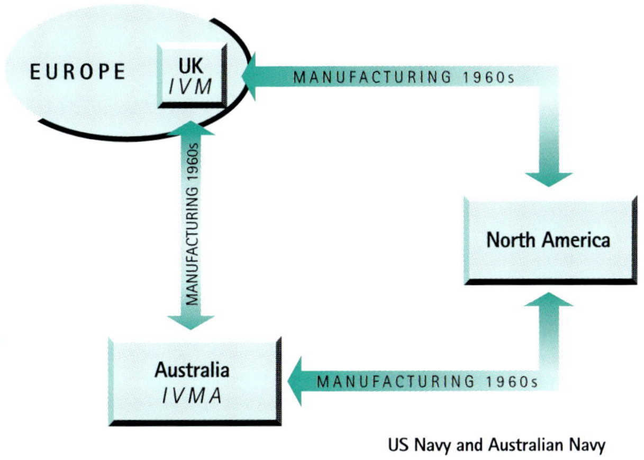

Figure 6 The diffusion of VM in manufacturing

The diffusion of VM in manufacturing from North America to the UK occurred during the 1960s. However, in Australia diffusion in manufacturing appears to have come through a number of different routes. First, practitioners from North America visited Australia in the manufacturing and government sectors. In addition, the Australian navy also adopted North American VM procedures within its own operations. Furthermore, the close Commonwealth ties between the UK and Australia during the 1960s also resulted in UK VM practice diffusing into Australian-based UK manufacturing subsidiaries.

Figure 7 summarises the institutional frameworks and diffusion processes in construction that have taken place between North America, the UK and Australia in construction.

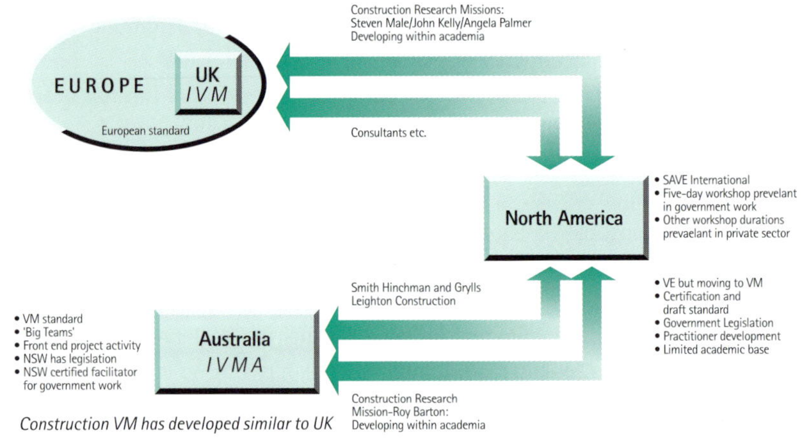

Figure 7 The institutional and cultural framework of construction VM

Figure 7 indicates that VM diffusion in parts of academia in Australia and the UK has occurred as a result of research missions. Interestingly, research study missions by construction academics from the UK and Australia concluded independently that North American style value engineering, with independent teams and five-day workshops, would not fit in with the host countries' cultures. Equally, other diffusion mechanisms have been in operation, with consultants, contractors and other types of organisations bringing VM/value engineering practices back to each host country. VM in the UK and Australia, as a consequence, has developed in a different manner in these countries than in the USA. Furthermore, VM in the UK and Australia has taken root in academia. However, in North America value engineering appears to have had a more limited impact on academia. Other issues to have emerged from the national and institutional comparisons are those associated with facilitator certification and also whether legislation hampers or hinders the deployment of VM throughout industry and government. These issues are discussed further in this research report.

5.3 Terminology and Definitions

There is considerable and confusing terminology being used surrounding the management of the value process internationally. The terms 'value management' (VM), 'value engineering' (VE) and 'value analysis' (VA) are still largely used interchangeably in the literature and it is not possible to isolate one area for investigation without reference to the others. Numerous names for the process or sub-processes are listed below and represent those in common use by VM organisations and institutions:

- VM value management
- VE value engineering
- VA value analysis
- VP value planning
- VR value review
- VM value methodology
- VMR value management reviews

The Building Research Establishment (BRE) states that it is not important what the process is called but it has opted to use VM as an all embracing term for any application. The Institution of Civil Engineers (ICE) uses VM as an all-embracing procedure. It also adopts the following terms as sub-processes within VM:

- value planning (VP), applied during the concept phase of a project
- value engineering (VE), applied during the definition phase of a project
- value reviewing (VR), applied at any point in the project life cycle to record the effectiveness of the value process which relates to the overall sequence of actions that lead to value improvements.

The Construction Industry Research and Information Association (CIRIA) uses VM as an all embracing term to describe a problem solving technique. However, it states that VM is aimed at strategy, early in the design, and that VE is concerned with achieving the project's objectives cost effectively. The Building Services Research and Information Association (BSRIA) states that VM addresses project objectives at the concept and feasibility stages and that VE addresses engineering issues in the design phases. This seems to concur with CIRIA. The HM Treasury Central Unit for Procurement (CUP)'s Guidance Note 54 makes reference to value management reviews (VMRs) throughout the life of a project.

The European Standard describes VM as a style of management encompassing various management tools of which VA is but one. It does however, state that VA is a tool used to tackle more tactical issues. VE is the term favoured historically by SAVE and since SAVE International has considerable influence internationally in construction the term is in common use around the globe. However, SAVE has recently renamed itself SAVE International and is now using the term 'value methodology' as an all embracing term. VA is the term embraced by the European Union Strategic Programme for Innovation and Technology Transfer (SPRINT) in publications that disseminate knowledge of the technique to all industry sectors across member states. The AS/NZS reiterates what BRE states in that VM is synonymous with VE and VA and can be applied any time during a project life cycle.

In analysing these disparate views it has been noted that whatever it is called, it is essentially the same process and that the name of the process has little to do with its application and could be considered a function of:

- the industry sector
- interpretation by particular professional bodies
- the desire to take ownership of a part or whole of the process by a range of different stakeholders with a vested interest in the subject
- marketing by various bodies and consultants
- historical development in respective countries
- harmonisation, in particular the European standard.

To summarise and conclude, a distinction between VM and VE is becoming evident in construction in the UK but there is inevitably still a large degree of overlap. VM in construction is increasingly being seen as the term to describe the total process of enhancing value for a client from a project through the phases of concept to operation and use. As a process, VM encompasses understanding and providing solutions to:

- *The 'business project'* - the business problem that the organisation needs to solve or address. The business project is unlikely to be defined in construction language but is expressed and communicated in the language of business. For example, the business project may involve reorganisation within the firm that may also require a new facility or refurbishment to a facility as options.

- *The 'technical project'* - the manner in which the business project is translated into the requirement for a built facility through design and construction.

VE is increasingly being viewed as a subset of the VM process, where the focus is on improving value in the design and construction stages of a project. Other terms are also being used in construction and there is merit in considering their use as a part of the overall VM process. For example, 'value planning' is the term being used to describe the front end of the VM process. VR is the term being used to describe the post-project audit stage where the completed facility is assessed in terms of its fitness-for-purpose against the original brief and design, although some authors, e.g. Norton and McElligot, are also using the term VA to describe this stage.

The research study has adopted the term VM as the most appropriate in construction to describe the process since it is applicable to a wide area that has as much to do with the business and the people involved in the process as a technical analysis of proposed projects.

5.4 VM and the Project Life Cycle

5.4.1 The Generic Value Management Process – Value Opportunities and Workshop Activity

The generic VM process can be broken down into two major areas:

- value opportunities throughout the project life cycle – possible intervention points
- the inputs to and outputs from the VM workshop – the process.

The research study has addressed each in turn.

Value opportunities – where to intervene in the project life cycle?
The documentation reviewed has revealed disparate views on this question. In fact, the use of different terminology, such as VA, VE, VM, VR, VMR, to project life cycle phases is made explicit in the different documents but also adds to the confusion. The following explains the views expressed in the various documents in turn:

BRE identifies four stages where VM may be carried out.

1.) Business strategy - indicating that VM be applied within a client's organisation prior to any decision to solve a problem using a built solution.
2.) Concept and development - relevant to the first phases of a project life cycle where cost reduction potential is high and resistance and cost to change are low. This represents a strategic approach to ensure project value reflects the client's previously established broad business goals.
3.) Design - to address issues of spaces, costs, time, etc..
4.) Construction.

ICE states that VP is applied at two stages and VE at one or more stages.

1.) VP1, early concept - to identify schemes capable of fulfilling the identified needs.
2.) VP2, end of concept - develop a brief reflecting that scheme which imparts appropriate value in light of the needs.
3.) VE1+ - post brief - aimed at translating the output from VP into a detailed design.

CIRIA approves the application of VM at four stages with a split between VM and VE. This split is dependent on whether, at the outset of the project, objectives are clear and shared by the stakeholders. If they are then VM encompasses the first stage and VE the other three, if not VM encompasses the first two stages and VE the other two, as set out below.

- Where the project objectives are clear and shared by stakeholders:
  - VM, concept - to ensure the project is thoroughly analysed
  - VE, feasibility - choose an appropriate scheme
  - VE, scheme design
  - VE, detailed design.
- Where the project objectives are unclear and not shared by the stakeholders:
  - VM, concept - to ensure the project is thoroughly analysed
  - VM, feasibility - choose an appropriate scheme
  - VE, scheme design
  - VE, detailed design.

BSRIA follows a very similar approach to that laid down by CIRIA.

The CUP's guidance note recommends six areas where it is appropriate to apply VM.

1.) Option appraisal and business case - the need for a project can be identified with key project constraints.
2.) Outline design - to define the project approach.
3.) Final sketch plan - to evaluate developing design proposals.
4.) Detail design - concurrent studies to review key design decisions as design progresses.
5.) Construction - to reduce construction costs or improve buildability.
6.) Handover - to learn lessons for future projects.

The draft European standard highlights the need for the application of VM to follow five project stages. It must be highlighted here that these stages appear to reflect an approach based on the manufacturing industry or, at best, an attempt to harmonise an approach suitable to both manufacturing and construction.

1.) Inception - establish needs and outcome.
2.) Concept - improve strategy.
3.) Feasibility - enhance viability.
4.) Implementation - maximise cost-effectiveness.
5.) Use - improve the product or process.

SAVE proposes applying VM as early in the design cycle as is feasible and indicates that value studies be conducted at two stages.

1.) Schematic stage.
2.) Design development (45%).

AS/NZS take a broader view (encompassing what seems to be the whole of the construction supply chain) and lists where typical VM studies may be conducted.

1.) Project concept
2.) Client brief
3.) Site selection
4.) Design proposals
5.) Material selection

6.) Construction programs
7.) Construction methods
8.) Facility management

The next section draws together information from fieldwork and the above review of guidance notes and standards to produce the benchmarked value opportunity points.

Benchmarked value opportunity points

Analysis of the above reveals considerable commonality and also a divergence of views in certain areas:

1.) Without exception, all approaches span that part of the project life cycle dealing with project concept, project briefing and design development.
2.) Almost all agree that the construction industry needs to apply VM as early in the project life cycle as possible. In some cases it is advocated that it should be as early as the pre-construction phase of a project where the VM process is used to establish solutions to an organisation's business problems. There may be no reference to construction as a solution at the outset.
3.) There is a difference of opinion on whether VM should be applied at a later stage than detail design, where cost reduction potential will be low, cost to implement changes to the project will be high and resistance to change will be high. Research being undertaken at Leeds University[11] on the use of VM/VE in the construction stage has indicated significant savings can still be produced at the construction stage. Based on this research, the premise, detailed in some of the documents of confining VM to during the project stages up to detailed design, is seen as deficient.
4.) One of the documents reviewed draws VM into facilities management and the analysis of construction supply chains

The application of the VM process follows the client's supply chain as it evolves during the project life cycle. This is represented schematically in Figure 8.

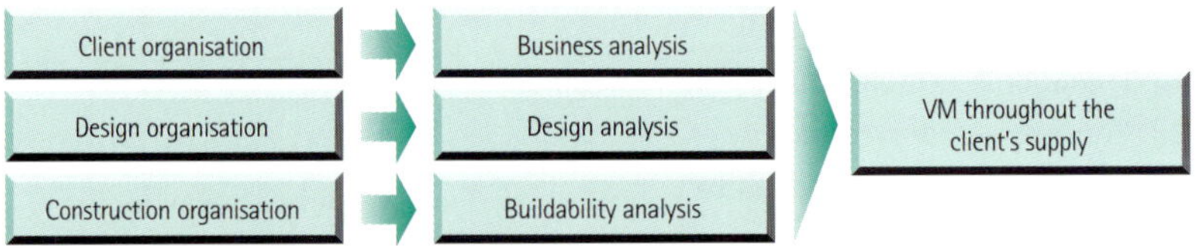

Figure 8 VM process and the client's supply chain

Summarising and incorporating findings from the research fieldwork shows there is a consensus that there are six points of convergence for VM opportunities in the project life cycle. These are:

1.) the pre-brief workshop, where the focus is on the business project
2.) the briefing workshop
3.) the concept design workshop
4.) the detail design workshop
5.) the Charette, it was also acknowledged that stages 1 and 2 could be combined as a review of the brief in an exercise termed a 'Charette'.
6.) the operational study undertaken at the construction stage.

While there is evidence to suggest VMRs are undertaken beyond the construction stage, they were not mentioned to any large extent in fieldwork

11. Standing N. *Value Engineering and the Role of the Contractor under different procurement routes.* Ongoing doctoral research, University of Leeds, in conjunction with Southern Water and Balfour Beatty Ltd.

discussions, as is the case also with facilities management. Therefore, there is insufficient empirical evidence to support the inclusion of facilities management and VMRs within the benchmarked value opportunity points. This does not, however, denigrate their importance. It merely indicates that at this point in time there may be few studies undertaken during these phases.

Each of the value opportunity points has been identified on the diagrams in Figure 9.

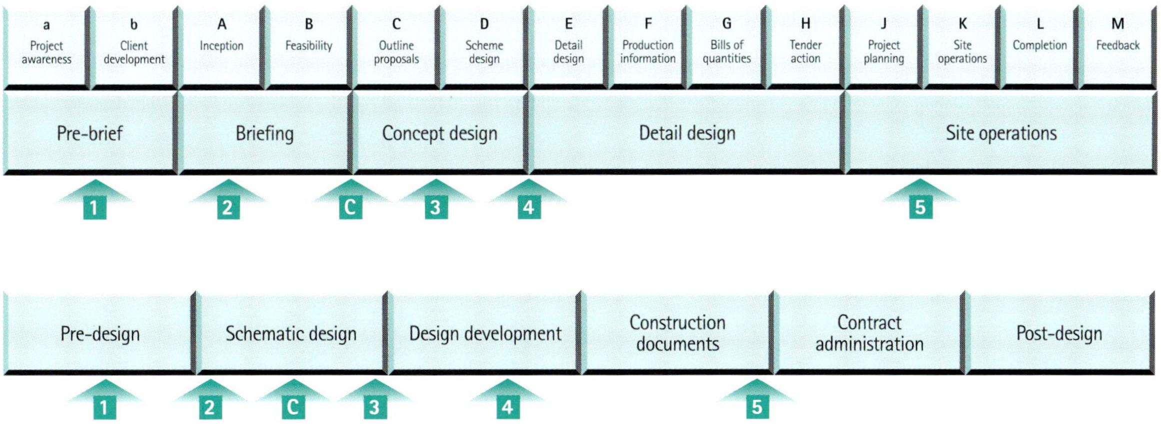

Figure 9 Value opportunities with reference to American Institute of Archtects and Royal Institute of British Architects processes.

The process at each value opportunity point

This section explores the issues surrounding the process of VM at the value opportunity points where the analysis of value takes place through the workshop process.

VM pre-requisites

The review of literature, guidance notes, standards and fieldwork developed a series of pre-requisites that should be in place to ensure a successful VM workshop outcome. Each is discussed in turn.

Senior management and participant support

Senior management within the client commissioning organisation and participant support for VM has been identified as a pre-requisite of success. Senior management sets the scene by endorsing VM and giving it their backing to ensure that everyone knows that ideas and suggestions arising from workshops will be implemented. This signals to participants that what they say and do will have an impact on the project.

Agreement to participate by all those attending the workshop is seen by many facilitators as fundamental to the success of the workshop. This may involve the facilitator interviewing individually or holding a pre-workshop meeting (orientation) with the aim of dispelling any nervousness of the process.

VM team characteristics

Almost all of the guidance notes and standards concur on this point – use an existing team and not an external, independent, separately appointed, shadow team for a VM study that works in parallel to the existing team.

The use of an independent external team is not thought to be beneficial due to the subsequent conflict this may cause with the existing design team when implementation of VM ideas arises. It was acknowledged internationally during fieldwork that the existing project team (the design or construction team) should be used as a base for selecting the VM workshop team. Supplementary experts can be brought into the workshop if required. It is important to note that these views are predominantly taken from Australia and the UK where studies frequently take place early in the project life cycle, i.e. between programming and outline sketch design. In the USA, on government contracts, the independent team is readily in evidence. However, on private sector projects, the use of existing teams is much more common. The argument for an independent team has been rehearsed in many of the texts on VM but in summary this revolves around first, bringing a fresh unbiased viewpoint to the project and second, providing the potential for technology transfer. However, this may be mitigated by increased problems for implementation, an atmosphere of confrontation and also increased project learning for the shadow team.

The advantages of using the existing team were quoted as being:

- costs are kept down.
- better ideas may crop up due to greater experience 'second' time around
- there is another chance to explore alternative options
- VM workshops are useful for team building and resolving conflicts that may exist within the existing team
- increased and more open communication within the team
- increased implementation.

There are a number of issues that arise from the above. First, the role of the facilitator is influenced by the nature of the team – existing versus independent. With an existing team the facilitator has to manage the VM process and assume a challenging role in order to assist the team members to 'break' from their traditional thinking on the project. On the other hand, for an independent team the facilitator has to be more oriented towards managing the VM process only.

In terms of team size and composition, BRE advises using the existing project team, relevant stakeholders and an experienced facilitator from within or outside the organisation. It does not, however, put a limit on the optimum, maximum or minimum number of participants in the workshop. ICE makes reference to the use of a 'value team' and states the following.

- Composition of the team should be flexible.
- It must include all relevant disciplines.
- It must be free-thinking and have company authorised decision makers present.
- Be led by the value manager.

However, ICE makes no mention of team size. CIRIA advises the use of a value manager who assists in developing and implementing a VM plan and may also facilitate the workshops. The value manager may be in-house or from an external organisation. Participants at their VM1 and VM2 points should be the client (or project sponsor), key project stakeholders, the project manager, design consultants and the value manager. The team should be multi-disciplinary and represent all interest groups. However, during VE, the team will consist of design team members and external parties which might comprise the facilitator, project manager, architect, structural engineer, services engineer, quantity surveyor, construction engineering advisor, general contractor, specialist contractor and facilities manager. The use of an external independent team is not thought to be beneficial due to the subsequent conflict this may cause with the existing design team. CIRIA does not make team size explicit. BSRIA reiterates the guidance given in the CIRIA document. The draft European standard emphasises that the team is a "mixed balance of individuals", adequately selected and trained. It also makes use of the term 'workgroups'.

SAVE International makes no reference to team size but does state that the value study team should represent the knowledge required to address the objectives of the study and that this will be project dependent. AS/NZS states a range of four to twenty plus for the number of personnel required in a VM workshop. This is dependent on the subject of the study. The team may comprise stakeholders, users, suppliers or multi-disciplinary specialists. A trained facilitator will also be in attendance to manage the VM process.

It can be concluded, therefore, that internationally there is a very strong tendency to use existing teams. Facilitators, in conjunction with the commissioning client, must use their skill and experience to decide whether an external independent team may be beneficial to the workshop and the project as a whole.

The analysis of the guidance notes and standards indicates no definitive answer to the optimum team size for VM workshops. This is also supported by fieldwork, although in Australia the research team encountered practitioners who specialised in 'big team' workshops at the front end of projects to reconcile different stakeholder interests. Roy Barton of the University of Canberra quoted workshop numbers of between thirty and forty during these studies.

The facilitator, in conjunction with the commissioning client, must be capable of making the decision on who should attend and why and the appropriate team size.

Types of participants

The empowerment of some of the participants of a VM workshop team to take decisions will influence the effectiveness of the workshop in terms of implementation of ideas. The term 'decision-taker' is used here to represent a person who has the authority to make and take decisions during the workshop. A 'decision-maker' can only provisionally endorse solutions and needs to refer to a higher authority in order for the solutions to be ratified and then implemented. It is proposed that the most effective workshop, in terms of implementation, is one where the client representative has the authority to take decisions and is a member of the VM workshop team. It is proposed that the least effective, in terms of implementation of ideas, is one where there is no client representative present and the team is fully independent.

It should be noted that the presence of a client decision taker may not be required for workshops conducted later in the project life cycle as the client's value system should already be encapsulated in the brief and in sketch designs. The VM process in this situation will be concerned with the performance of elements and components. However, it would be necessary to bring a client decision taker on board if no VM workshops had been carried out on the project previously.

What is a facilitator?

It is widely iterated in the documentation and also supported by fieldwork in the benchmarking study that the facilitator should possess the following attributes:

- experience
- training in the VM process and its relevant tools and techniques
- independence, drawn from either in-house personnel but not directly involved with the project, or external to the organisation
- competence in a variety of management skills related to human dynamics and the management of teams
- appointed by the client, or project sponsor, for a specific purpose
- possess leadership qualities.

Some interviewees view the facilitator as needing to have relevant skills to the project under consideration. In international fieldwork, experienced facilitators were cited by the majority of benchmarking partners as being paramount for a successful VM workshop. A significant number of interviewees also went much further in this assertion, especially those in commissioning client organisations of VM studies. They indicated a growing loyalty to a particular facilitator by clients and not to the consultancy organisation from which the facilitator is commissioned. This may well set an upper limit on the number of potential facilitators available for VM studies. The research evidence clearly supports the view that good facilitators are a scarce resource. This is taken further in the section dealing with the professional development of VM.

The facilitator is, in effect, a VM process manager and acts as team co-ordinator, guiding the team through the tools and techniques best matched to the project under investigation. Key skills were highlighted as being:

- setting the workshop agenda
- managing the process of the workshop, including:
  - dealing with any hidden agendas of participants
  - recognising individuals and their contribution
  - questioning and summarising
  - providing direction and a sense of common purpose
  - sensing interpersonal relationships within the team
  - sensing the climate of the workshop
  - synthesising and integrating information during the workshop
  - intervening in the workshop process when necessary
- communicating verbally and producing the VM report.

A related and unresolved issue from the fieldwork is the use of approved/accredited facilitators. This may potentially constitute the birth of a new profession. This is explored further in a later section.

Externally appointed or internal facilitators

In terms of the debate over external versus internal facilitator from within the client organisation (or alternatively those associated with the project is some form or another), the consensus is that an external facilitator is used. The definition of external is someone who is not explicitly involved with the project being analysed. External facilitators are considered to be more effective than internal ones due to the fact that they are likely to pose probing and challenging questions without fear of sounding unknowledgeable. An external facilitator also has no vested interest or hidden agendas with regard to the project.

The number of facilitators – single versus dual facilitation

A further unresolved issue from the documentation studied is the number of facilitators required to manage the VM workshop efficiently. The fieldwork indicated clearly that dual facilitation (two facilitators as opposed to one facilitator or one facilitator with a recorder) has emerged as the preferred option due to the mutual help the facilitators can give each other and the variation it gives the participants. Two facilitators alternating can keep the momentum going in a workshop to a higher degree than someone working on their own. Some stated that it is also possible to work longer days without sacrificing quality. However, while this was the preferred approach it was acknowledged that the economics of the market place tended to drive facilitators to quote for a single facilitator option or sometimes one facilitator with a recorder. It raises the question as to whether clients are, in fact, getting value-for-money from the VM process by forcing prices down through the market place and the consequent impact on the quality of service provision.

Large and small team facilitation

In Australia, as mentioned earlier, there are VM practitioners who tend to specialise in large team facilitation, often with team sizes in the range of 30-40 participants. Facilitating such large numbers of team members will require a different approach to the process of managing a workshop. It will normally involve forming smaller working groups, often operating in parallel on different tasks or on the same task. The VM facilitator(s)' task may well involve:

- facilitating multiple teams
- facilitating multiple facilitators selected from within the smaller working groups - the small working group facilitator may have some formal skills in facilitating already developed, or may have no formal facilitation skills but a natural ability for the task
- the ability to synthesise information very quickly in order to keep workshop momentum going and the agenda on track.

The facilitation of large teams is likely to require an enhanced skill base than that for facilitating much smaller teams in the order of 6-10 participants.

Pre-workshop issues

BRE advises the use of a pre-workshop meeting between the facilitator and the stakeholders to define the workshop objectives. ICE states that a preparation stage should be implemented prior to the workshop to agree:

- terms of reference and authority of the VM study and the participants
- objectives of the project, components or elements
- timescale for the workshop
- workshop budget
- workshop resources required and their availability
- date and location of the workshop
- the owner's 'value criteria', if available.

The value manager must also:

- establish the team
- gather, list and distribute information
- set the workshop agenda.

The ICE makes no reference to preparation required for the VP exercises. CIRIA also advocates a general preparation stage to:

- select and brief value manager
- provide an initial statement of need
- identify stakeholders

- organise and participate in the workshop
- agree the debrief, implementation and feedback process.

SAVE International identifies a rigorous pre-study system of activities:
- collect user/customer attitudes via focus groups or market surveys
- gather a complete data file
- determine evaluation factors
- determine the scope of the study
- identify models of cost, energy, etc.
- wrap up this phase by confirming study schedule, personnel required, location, etc.

AS/NZS again reiterates the need for a preparation phase to be completed prior to the workshop commencement, similar to the previous documentation.

Interviewees in the fieldwork unanimously agreed that preparation for a workshop by the facilitator has been seen as critical for a successful workshop. This is often undertaken by interview, file search and documentation analysis, facility walk through and pre-workshop orientation meetings.

The study concluded that a pre-workshop study information stage should be included as part of the VM workshop process.

Workshop issues

The job-plan and its variants

There are very few disparate views expressed in the documentation analysed on how the workshop should be executed. All make reference to a 'job-plan' or 'work-plan' which is based on the Miles' original 'job-plan'.

Functional analysis and objectives hierarchies

There is one fundamental issue, highlighted in the documentation, where views diverge. This is the use of analysis by function or analysis by objectives. The origins of VM, as proposed by Miles, are rooted in the use of functions and functional analysis. BRE uses analysis by function. ICE uses objectives and value trees during VP as a way of assessing the 'purpose' of the project but then advocates the use of analysis by function and FAST diagrams later in the project life cycle when applying VE to help translate the output from VP into detailed design. CIRIA and BSRIA also use objectives and value hierarchies during the value workshops early in the project life cycle but state and use function during the application of VE later in the project life cycle. The draft European standard, SAVE International and AS/NZS standards use analysis by function throughout the project life cycle and in the associated VM workshops.

The fieldwork internationally demonstrated functional analysis as the preferred analytical approach in workshops. This is consistent for both manufacturing and construction. It is the approach adopted in this study.

The divergence of views between functional analysis and objectives hierarchies has only proved to be an issue within the UK construction industry. There are similarities between the function logic diagrams inherent within the structuring of functional analysis and that used in creating objectives hierarchies. However, reference to the discipline of psychology and the theories inherent within personal construct psychology developed by Kelly[12] can easily reconcile these differing perspectives. At a fundamental level functional

12. Kelly G. *The Psychology of Personal Constructs.* Vols 1 and 2. Norton, New York, 1955. Bannister D. and Fransella F. *Inquiring Man: The Theory of Personal Contructs.* Penguin Books, Baltimore, 1977.

analysis and objectives hierarchies both use natural language to structure a value-oriented problem in a workshop environment around a hierarchical diagramming technique. The debate over functional analysis or objectives hierarchies becomes unproductive when looked at in the context of their respective contributions to problem solving in a workshop environment.

Workshop environment

The likelihood of a VM workshop being successful is enhanced if an isolated environment is provided for the participants where there are no interruptions from outside agents. Advantages cited include:

- it focuses the team on 'the project'
- gestation occurs during the workshop process
- it commits the team
- militates against partial attendance
- continuity is ensured.

It is important to weigh up the cost of providing the isolated environment to maximise the outcomes against the apparent cost. An isolated environment, for example in a hotel, does focus the team totally on the project under investigation in a 'pressure cooker' environment. Residential studies also stimulate team building but workshop costs are higher than non-residential situations.

Post-workshop issues

Implementation of workshop proposals is the 'acid test' for VM. It is surprising that the guidance notes and standards do not put a larger emphasis on what must be accomplished on completion of the workshop. However, ICE does make reference to an implementation/follow-up stage for VP and VE with the explicit objective of ensuring the implementation of workshop proposals. Post-workshop reports are highlighted as an integral part of ensuring successful implementation after VE. CIRIA and BSRIA also highlight post-workshop reports as an integral part of ensuring the successful implementation of workshop proposals. BSRIA provides a pro-forma of a typical VE report in its appendix. SAVE International recommends assigning tasks to the workshop team members in order to ensure the implementation of workshop proposals.

Implementation of proposals was consistently highlighted during the fieldwork for the research as being critical to the success of VM. Implementation was seen as a significant area of weakness for VM. In essence, the unsuccessful implementation of proposals, accepted as adding value to a project during a VM workshop, should be considered as a failure on behalf of VM to deliver value.

The study concluded that a post-workshop study stage of implementation should be included as part of the VM workshop process.

The Benchmarked Value Management Process

The above discussion is now drawn together into a statement of the benchmarked VM process for workshops.

The VM process has a series of pre-requisites that must be satisfied to ensure a smooth running workshop. These mainly relate to the involvement of people and the venue for the workshop and include:

- agreement to participate by all parties involved in the workshop
- senior management support for VM
- an experienced and independent facilitator(s)
- an appropriate team skill mix
- the presence of client decision taker(s)
- an isolated workshop environment.

However, these are necessary but not sufficient to guarantee a successful outcome for the client commissioning a VM study. A pre-workshop phase is required as part of the VM process. The pre-workshop phase should be concerned with:

- gathering information
- eliciting views and opinions of key stakeholders and project specific participants
- helping to define workshop scope
- identifying project constraints
- contextualising the value problem within both the 'business' and project delivery process
- briefing the workshop team and facilitator prior to the workshop.

It may be necessary in some situations to include a pre-workshop meeting to address the objectives of a large number of project stakeholders at the concept stage of a large, complex project. This may be inappropriate in other situations and is part of the facilitator's job to decide on this course of action.

The facilitator will have a leading role in determine team size and composition. The lack of clear empirical evidence on team size is due to:

- the nature of the problem under investigation at any given workshop
- where in the project life cycle the VM workshop is being conducted - this will determine how many relevant stakeholders are required and the range of expertise required
- the resources available for the study.

The most accurate definition of team composition and the required numbers for any particular workshop will be a function of the knowledge required to solve the problem and the necessary views and opinions of those directly influenced by the workshop outcome. This is related to project size, value and complexity and it is the task of the facilitator to propose the team composition and size.

Workshop length is a function of various factors that the facilitator must be able to analyse and have a leading role in determine the optimum length of study required. However, in terms of workshop duration, one to two-day workshops are the norm in the UK and Australia. There appears to be a wide variation in duration in the US, from half a day, one day and up to five days encountered. Another issue to have emerged in the study is where the workshop should be residential and off-site or on-site. Practice favours residential in the UK and North America but not Australia. The residential and off-site workshop has the advantage of focusing the team on the project, it is intense but gestation occurs during workshop process. Equally, it commits the team and militates against partial attendance if it is on-site. Again, it is the facilitator's task to propose workshop duration and location.

A workshop should not be considered a success unless its outputs/proposals have been implemented on the project. Post-workshop reports are extremely useful in achieving this if they impart an auditable chain of responsibility on the workshop team members to ensure implementation.

The VM process is summarised at each of the six value opportunities in Figure 10.

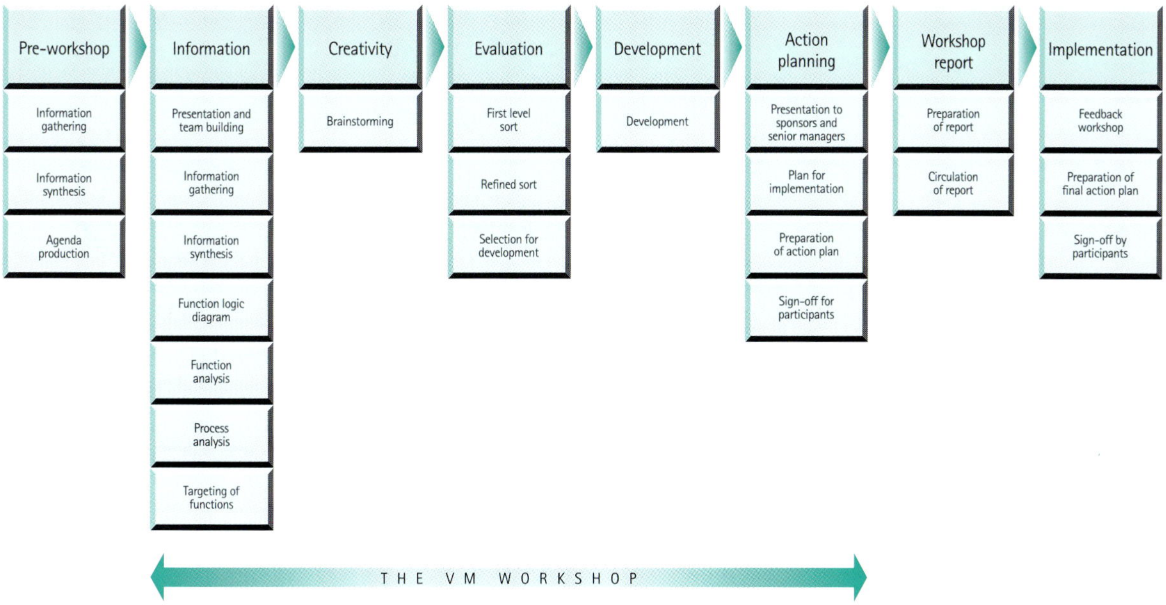

Figure 10 The generic VM process

The next section deals with workshop structures.

5.5 Individual Value Opportunity Workshop Structures

5.5.1 Introduction

As stated above, in construction there is consensus on the six opportunities where the VM process could be employed to achieve maximum effect on any project during its life cycle. These are:

- the pre-brief workshop
- the briefing workshop
- the concept design workshop
- the Charette
- the detail design workshop
- the operational study.

Each of the workshop structures indicated above is described in more detail in the accompanying *Good practice framework for clients and practitioners*. The workshop structures represent a synthesis and drawing together of information identified in the literature and during fieldwork.

5.5.2 Pre-brief workshop – the strategic brief

The strategic brief sets out the broad scope and purpose of the project in its business context and its key parameters, including the overall budget and schedule. The strategic brief should provide an output specification that explains in clear terms what is expected of the project, and the organisation's value criteria. The project at this stage exists as 'the investment of resource for return' and does not necessarily imply construction.

The primary purpose of the strategic brief is to structure information in a clear, unambiguous form to permit a 'decision to build,' if this is the preferred option. The decision can be taken in light of all of the pertinent facts, which include the output of brainstorming sessions. The strategic brief ought to include:

- the mission statement for the business project
- the project context
- project organisational structure
- overall scope and purpose of the project
- a global capital expenditure budget and cash flow constraints
- targets and constraints on operating expenditure and other whole-life costs
- what is expected in response to the brief
- how the success of the project will be measured and the client value criteria
- statutory requirements.

The workshop is likely to last between one-half and one day and involves a facilitator and senior representatives of the client organisation only. Since it is important to hear the views of all stakeholders of the project the numbers attending this style of workshop tend to be large, between 10 and 30, possibly more. This requires innovative team management and may involve splitting into subgroups for short periods to study assigned tasks.

It was stressed during the benchmarking interviews that the facilitator is required to be well prepared for these studies. This may necessitate extensive pre-study interviews and the preparation of 'information packs' for all participants. Indicative tasks at the workshop include:

- presentation and team building
- checklist or brainstorm issues, group and theme
- stakeholder analysis
- strategic time line
- project driver analysis
- time/cost/quality analysis
- function analysis
- function logic diagram
- brainstorming alternatives
- evaluation and development
- presentations
- planning for implementation
- preparation of action plan
- sign off by participants
- preparation and circulation of report by facilitator.

5.5.3 Briefing Workshop - The Project Brief

The project brief (or programme) re-states the strategic brief and converts it into a performance specification in constructional terms, detailing:

- the aims of the design, including prioritised project objectives
- the site, including details of accessibility and planning
- the functions and activities of the client
- the structure of the client organisation
- key targets for quality, time and cost, including milestones for decisions
- schedule, including phasing
- the size and configuration of the facilities
- statements on size and capacity requirements and functions to be accommodated including outline specifications of general and specific areas

- environmental policy, including energy
- internal and external environmental requirements including options for environmental delivery and control
- technology to be incorporated or accommodated, including equipment, services, information technology (IT)
- quality requirements for design, materials, construction and long-term maintenance
- servicing and maintenance options and specification implications.
- a budget for all elements
- the procurement process
- the project execution plan
- method for assessing and managing risks and validating design proposals.

The primary purpose of the project brief is to structure information in a form that can be readily understood by the design professionals and the contractor. The project brief should overtly represent the client's value system as identified in the strategic brief.

The workshop is likely to last between one and two days and involves a facilitator, the client's project sponsor(s), who is likely to be a senior representative of the client organisation, together with the design team. The views of all stakeholders of the project have been recorded in the strategic brief and therefore the numbers attending this workshop tend to be between 10 and 20.

Indicative pre-workshop tasks include:
- conducting interviews with client using a checklist
- document analysis
- site tour
- use of questionnaires
- post-occupancy evaluation (POE) reports of similar buildings
- facilities walk-through of similar buildings.

Workshop tasks include:
- presentation and team building
- review of function logic diagram, stakeholder analysis, time/cost/quality analysis and strategic time line
- REDReSS (re-organisation, expansion of the organisation, disposal strategy, refurbishment and maintenance strategy, safety and security)
- SWOT (strengths, weaknesses, opportunities, threats) analysis of the design to date
- checklist or brainstorm issues, group and theme
- process flow charting
- functional space analysis resulting in identification of, for example, size, shape, quality, heat, light, acoustic attributes, IT requirements, future demands.
- spatial adjacency
- brainstorming alternatives
- evaluation and development
- making presentations
- planning for implementation (the project brief)
- preparation of action plan
- sign-off by participants
- preparation and circulation of brief and action plan as a report by the facilitator.

5.5.4 Brief Review Workshop – the Charette.

This is an alternative to the preparation of the brief through the VM process and involves the review of an existing brief commonly prepared by a client, project manager or architect. In the Charette, all of the stages of the strategic brief and project brief workshops are combined into a single workshop lasting between one and three days. This style of workshop is common in Australia and the UK. One UK value manager stated that the vast majority of all workshops facilitated were of the Charette type. One Canadian client stated that a Charette was the best way of briefing the design team and ensuring that the team fully understood the client requirements.

5.5.5 Concept Design Workshop – Outline Sketch Design (OSD)

The concept design workshop is a value review of the initial plans, elevations, sections, specification, and cost plan of the proposed facility using the signed-off concept brief as a reference point. Its primary aim is to review and improve the concept design. The concept design ought to include:

- a statement of the design direction
- the site layout and access, identifying ground conditions and planning constraints
- a detailed cost plan and schedule of activities
- dimensioned plans, elevations and sections
- an outline specification for environmental systems
- the risks and a risk management strategy
- the procurement plan
- the project execution plan with key milestones
- performance measures.

The OSD workshop is likely to be of two to three days duration and involve between seven to twelve participants, comprising the design team and the client project sponsor/manager. Because the client's value system is defined in the brief/programme it is no longer necessary to involve extensive client representation. Workshop tasks include:

- presentation and team building
- review of function logic diagram and concept brief, particularly functional space requirements and adjacency
- allocating costs from the cost plan to functions and spaces
- highlighting mismatches and risks
- brainstorming solutions
- evaluation and development of solutions
- presentations
- plan for implementation of preferred solutions
- preparation of action plan
- sign-off by participants
- facilitator to prepare list of solutions and action plan as a report.

5.5.6 Detailed Design Workshop – Final Sketch Design (FSD)

Once the client project manager has 'signed-off' the concept design the project team should begin the development of the detailed design and specification of the performance requirements for elements of the facility. The detailed design workshop should freeze as much of the design as possible, defining and detailing components of the construction work. It should identify risks associated with the project and outline proposed action if they arise, assess the quality requirements and define how success will be measured. The detail design workshop ought to include:

- a statement of scheme design
- location of facility and infrastructure requirements, planning approvals and other detailed permissions agreed
- dimensions of spaces and elements provided with some description of components
- performance specifications for environmental systems and services
- the cost plan
- proposals for the maintenance and management of the completed facility
- the project execution plan with key milestones and targets
- performance measures.

The detailed design workshop is likely to be of two to three days duration and involve between seven to twelve participants, comprising the design team, the client project sponsor/manager and the client facilities manager.

Workshop tasks include:

- presentation and team building
- review function logic diagram, concept brief and concept design
- identify element and component functions from detailed design
- allocate costs from the cost plan to element and component functions
- undertake life cycle costing studies
- highlight mismatches and risks
- brainstorming solution
- evaluation and development of solutions
- presentations
- plan for implementation of preferred solutions
- preparation of action plan
- sign-off by participants
- facilitator to prepare list of solutions and action plan as a report.

5.5.7 The Operational Study

The operational study, undertaken during the construction phase, is to convert the design into components and construction operational sequences. It is likely to be of one or more days duration, with six to ten participants including production planning, purchasing, project management, supplier and/or subcontractors. Of all the workshops described, this is likely to be the most variable in terms of its structure as the focus could be the whole project, an individual work package or a component.

Workshop information is likely to include:

- contractor's contract strategy
- production drawings and contract programme
- histograms from tender information
- major work packages
- change management procedures.

Workshop tasks are likely to include:

- interviews with key contractors' personnel, client representatives and design team members
- presentation and team building
- reviewing the contract programme
- supply chain analysis
- functional specification analysis

- component analysis
- Pareto analysis - histogram of cost
- pinch point and committal point analysis
- detailed time/cost/quality analysis
- life cycle costing
- production plans
- highlighting mismatches and brainstorming alternatives
- evaluation and development decision matrices
- presentations
- plan for implementation
- preparation of action plan
- sign-off by participants
- the facilitator prepares and circulates the study report.

The main output from the operational workshop is a validated production programme.

5.5.8 The Implementation Workshop

The aim of this workshop is to ensure that recommendations set out are being implemented. The workshop will follow the concept to detailed design workshops. During benchmarking, examples of high implementation rates were cited where an implementation workshop was held. An indicative duration was given of between one half to one day. The potential participants were as in the workshop being followed up plus a representative of client's senior management. The later is considered necessary to add to the credibility of the exercise.

Workshop tasks include:

- recommendations and findings of the workshop being followed up and summarised in a presentation by the participants
- workshop participants make presentations on progress within their allotted duties from the workshop action plan, implementation issues are included together with problems encountered
- brainstorming solutions to problems encountered or recommendations that have run into difficulties is conducted, these are reviewed and refined
- brainstorming of new implementation issues is conducted, together with reviewing and refining
- recommendations and a final action plan is developed
- sign-off by participants.

5.5.9 Tools and Techniques

The techniques used by facilitators during the various workshops outlined above were too numerous to describe in this report. However, all of the techniques have been captured and are described in the accompanying *Good practice framework for clients and practitioners.*

The next section deals with issues associated with the continued professional development of VM in the UK.

5.6 The Market and Professional Development of VM

5.6.1 The Influences of the Market Place on the Provision of VM Services

The role of clients

The study has identified the important role of commissioning clients in diffusing VM practice. It is the large, knowledgeable and regular procuring clients that are using VM on a frequent basis. They will often have sophisticated methods for choosing facilitators. VM is likely to be embedded in a project delivery manual as a formal review procedure. However, in the UK a different VM research study involving the applicants[13] has revealed that small and medium sized, less knowledgeable clients are often difficult to identify. The diffusion to them of VM practice and procedures is therefore hampered. The findings concerning large, regular procuring clients is also repeated in North America and Australia. This indicates quiet clearly that the diffusion of VM is being spearheaded very much by experienced clients and there is likely to be a cascading or trickle down effect to the irregular procuring client domain.

Within client organisations regularly procuring VM services, the study has confirmed internationally there is a danger of VM falling into the trap of 'executive boredom'. To work effectively, the VM process needs executive representation at workshops, as identified above as, one of the important pre-requisites. Empirical evidence has indicated that senior managers may well become bored with constant attendance at these events. There is also a danger that VM may be seen as another management fad. Equally, VM may stray easily into a 'going through the hoops' procedure and hence lose the creativity inherent within the process. There is also pressure from commissioning clients to reduce workshop time to a minimum. Equally, the benchmarking study also came across many examples where VM is a mandatory requirement by various commissioning bodies. This may initially assist VM dissemination but can lead to box ticking. This is addressed further below in the section dealing with qualifications, standards and legislation

Employing the firm or employing the person

The majority of VM practitioners see the provision of VM services as facilitating the application of VM to a project. The nature of the service therefore becomes one of short, intensive interventions at specific value opportunity points in the life of the project. Some organisations, particularly manufacturing, process and utility companies have so many projects on stream at any one point in time that it is effective to employ VM facilitators for full time VM work. In contrast, other organisations have infrequent demands for VM and therefore specialist consultants provide this service.

In Australia, and latterly in the UK, main contractors and also surveying and project management organisations have initially employed specialist VM consultants to the construction industry. However, clients identify certain facilitators as being suitable for particular types of projects and their choice of facilitator tends to be based upon individuals rather than the employing consultancy practice. In Australia, this has resulted in a market place characterised by four distinct types of VM practitioners.

1.) As one interviewee termed it, there is a 'cottage industry' of sole (or perhaps two) VM practitioners who work from a low-cost office at home or elsewhere on a full-time basis.
2.) The 'cottage industry' practitioner working in association with a larger consultancy from within the latter's offices but not directly employed by them. There will be some form of financial arrangement or structure in place for the use of office space and the use of the umbrella consultancy name.
3.) The VM consultancy with a number of practitioners who deliver VM as their core service.

13. Building Research Establishment. Value from construction - getting started in value management.

4.) A VM practitioner employed directly by a consultancy or other type of organisation whose core activity is not VM. The VM practitioner may or may not be involved in the provision of a VM service on a full-time basis.

Examples of 1, 3 and 4 above have also been found in the UK and USA. It is worth noting that in the UK VM developed initially along the lines of Type 1. and to a greater extent with both of Type 3 and variants of Type 4.

Building on Types 1 to 4 above and integrating this with the empirical evidence from the fieldwork, it is possible to discern a typology of VM practitioners that have different tensions and pressures stemming from the market place. Type 1, the pure 'cottage industry' practitioner, is able to compete very effectively on price with a low-overhead structure. Pressures stem from the need to maintain volume of work and also resourcing, which becomes an issue when work expands. This type of practitioner may operate within a network of VM practitioners that militates against the resourcing problem.

Type 2 is similar to Type 1 but may benefit from the advantages of marketing under an umbrella consultancy organisation. Equally, they may also have the additional benefit of being subcontracted into consultancy work for the umbrella organisation that may wish to offer or use VM on its own projects.

Type 3 requires volume of work or a unique specialist service permitting high fees to be charged or some combination of both to sustain its business over the longer term. It may be susceptible to internal tensions due to clients preferring particular facilitators to those also available from the employing consultancy organisation. The resultant '*prima donna*' centrifugal force with a consequent possibility of spin-off competing 'cottage industry' practitioners who take their clientele with them may be too great for the consultancy organisation to retain internal cohesion. It may also have resourcing and growth implications.

Type 4 comprises two subtypes. One is essentially a Type 1 practitioner but with the benefit of full-time employment within an organisation. The main problem facing this type of practitioner is the ability to sustain full-time employment from VM to the satisfaction of the employing organisation. The second subtype is the practitioner whose sole activity is not VM. The advantage of this is that the organisation and the individual are not reliant on VM to sustain business income. However, the main disadvantages may be demonstrating a lack of commitment to VM to potential clients or an inability to keep abreast of VM current practice and issues.

The preceding tensions are already surfacing in the supply side of the UK VM 'industry'. A series of propositions can be put forward for the structuring of VM in the UK. Manufacturing, process and utility companies due to their volume of work are in a position to build up in-house VM organisations who will track and value analyse projects. In construction, the larger contractors are also likely to find an in-house VM organisation attractive for the same 'competitive advantage' reasons as manufacturers. Outside of these organisations the UK is likely to see the rise of the VM 'cottage industry', in one form or another, and large consultancy practices may find difficulty in sustaining a VM capability without formal or informal links to those in the 'cottage industry'.

There are a series of implications stemming from the foregoing for the continued professional development of VM in the UK – the subject of a later section.

5.6.2 Policy Issues: Qualifications, Standards and Legislation

The important policy issues to have emerged from the research for consideration by UK government departments, clients or professional associations are associated with certification of facilitators, legislating for the use of VM and the development of standards.

In terms of the certification of facilitators, the important issue is that good facilitators are seen to be in short supply. They may well be a self-selecting group of individuals that have a natural ability that is then developed through experience and, perhaps, training. The use, therefore, of only certified facilitators on VM studies may well restrict the overall deployment of the methodology and philosophy within industry and academia. On the other hand, certification does provide the client or end user of the service with peace of mind that they are using a qualified individual. The question still remains, however, of which is the appropriate institutional framework that provides the best mechanism for certification.

With respect to legislating for the use of VM, there is no doubt that it assists in the deployment of VM in instances where it is a mandatory requirement. However, the counter argument is that this may eventually lead to a 'tick the box' activity. This militates against creativity and improvements that the VM process can bring to both project and product delivery.

A number of countries have standards for VM, including Austria, Australia and New Zealand, France, Germany, Hungary, India, Poland and USA. There is a European standard currently dealing with definitions and a more full standard is to follow, being the aim of the SPRINT programme described earlier. The debate on "why have a standard?" is interesting since there is no British standard for, for example, architecture or quantity surveying. However, 'architect' is a legally protected name guaranteeing the competence of the holder of that title. While quantity surveying is not likewise protected it is recognised that a chartered quantity surveyor knows sufficient about the measurement and value of construction to act competently with the protection of professional indemnity insurance.

It can be deduced that a standard is necessary because the service of VM is sufficiently vague in the minds of those who commission VM. This issue of definition is dealt with by the USA public sector by making reference to a standard 40-hour workshop. In Australia, the New South Wales Government's *Total Asset Management Manual* defines the service as well as the rules. However, the development of standards is a double-edged sword. In the first instance it acts as a benchmark or comparator against which practice can be judged for adequacy. Standards are also useful in defining, reinforcing or clarifying terminology for wider acceptance. However, by its nature, VM is a creative and change-oriented process and the use of standards may proceduralise practice, hamper innovation and hence deter experimentation as a vital ingredient of methodological development. The same argument equally applies in this instance to the certification of facilitators, where some form of standard procedure is required to determine if an individual is able to become an approved facilitator.

Certifications, standards and legislation for VM are also related to the professional development of VM – the subject of the next section.

5.6.3 The Emergence of a New Profession – the VM Facilitator?

There are a number of questions that need to be addressed in terms of the professional development of VM in the UK. The fieldwork for the study has identified and clarified these. This analysis will be undertaken in four parts by investigating:

- the defining characteristics of professions
- the collective knowledge base of VM and whether this is founded on theory
- the application of that theoretical knowledge base by practitioners
- education and training.

A series of conclusions are reached on the service delivery of VM within an occupational and professional context.

What is a profession?[14]

The term 'profession' is a much overused and confusing word to describe a particular occupational phenomenon in the business and social world. There is no doubt that professions are held in high esteem by society in general and many occupations seek to attain professional status. A characteristic of the growth of a profession is the demand for the service. As the demand for a service grows so the service becomes more formal and recognisable in terms of its skills, inputs and outputs. The boundaries of the knowledge base become established (for better or worse). A further feature of the growth of a profession is the monopoly, power and influence of the profession in society and the price that the profession can charge for its services.

Professions can be viewed as a special form of occupation and as the latter become more 'professional' they are said to be professionalising. An occupation is defined as:[15]

> "the social role performed by adult members of society that directly and/or indirectly yields social and financial consequences and that constitutes a major focus in the life of an adult"

An occupation is also related to the similarity of activities carried out within the general scheme of the division of labour within society. In looking at VM from an occupational and professional perspective, in construction the delivery of the service is focused on ensuring value-for-money for clients in the project delivery process, from inception to handover, commissioning and into use. It achieves this through a process that culminates in a workshop format at various potential value opportunity points in the project by bringing together different experts under the guidance of a VM facilitator. Therefore, a VM service can be described as a further division of labour within society and the business world. This is consistent with the view of an occupation. VM also has a set of guiding principles about when it may be appropriately used and an associated set of tools and techniques. However, is this sufficient to describe it as an emerging profession?.

Professions are a form of occupational control and, at best, they are a unique product of the division of labour in society involving highly specialised skills. Professions are also involved in social, societal and market based sets of relationships from which stems their power, prestige, status and potential for high incomes. Professionalism is a term associated with professions. It is an occupational ideology used by those that are traditionally called 'professions'

14. The Section is based on: Male S.P. *A Critical Investigation of Professionalism in Quantity Surveying*. PhD thesis, Heriot-Watt University, 1984; and Spiteri J. *The Strategic Management of Architectural and Quantity Surveying Practices*, ongoing doctoral studies, University of Leeds.

15. Hall. *Occupations and the Social Structure*. 1969.

such as medicine and the law. Equally, it is used by some occupations seeking to gain prestige and status. In order for an occupation to be seen and accepted as a profession requires public recognition. Professionalism, as an occupational ideology, is essentially a strategy used by an occupation to seek, gain or retain occupational power through the 'credentials' offered by its related professional associations. However, there are many types of professional associations, from those that are essentially trade associations representing the interests of their members, to learned societies, through to associations concerned with ensuring individuals are fit to practice. Some professional associations combine a number of these functions.

Within the context of professions, credentials are usually conferred by professional associations and the most important for professional behaviour is the qualifying association. The qualifying association exercises control over the occupation. It does so by testing competence and through the consolidation and development of the profession's knowledge base. Qualifying associations also provide control over recruitment and training and they are also involved in the educational process. In conjunction with the occupational control exercised by qualifying associations, knowledge is an important power base for professions resting on expertise. Self-regulation is also normally a profession's best claim to status and many of the established professions are not only self-regulating but also have codes of conduct to regulate professional behaviour.

From the preceding and using a number of models of professions and professional behaviour, there are a number of different ways of viewing professions:

- *The trait approach.* This attempts to identify common attributes that separate professions from other occupations. Typical traits of 'professions' are:
  - self-organisation through professional associations
  - adherence to a code of conduct
  - the requirement of training and education, normally to degree level for prospective entrants
  - a complex occupation with a knowledge base requiring the exercise of specialist skills
  - skills based on theoretical knowledge
  - a sense of vocation and belief in altruistic service.

  Occupations can possess more of less of the above traits in their progress to becoming a profession.

- *The market closure perspective.* This approach argues that through an occupational strategy of professionalism, using the credential possessed by members of the profession and awarded by their professional associations, professions are able to exert market closure and control the prospective number of candidates entering. This enhances the market value of the profession through the forces of supply and demand.
- Occupations can possess a combination of *cruciality* and *mystique*. An occupation possesses cruciality if it is perceived as vitally necessary for prosperity or survival to 'significant others' using the occupation. Mystique arises from cruciality. When a person or organisation seeks out a practitioner seen as crucial for solving a problem then the extent to which that practitioner is perceived as possessing extensive knowledge on the subject concerned, compared to the lay person, creates an aura of mystery or mystique and hence and authority relationship on the part of the

practitioner. The greater the cruciality and mystique of an occupation the greater the status of an occupation as a profession.

- *Occupational segmentation.* Professions are major segments of practitioners who are organised coalitions around a particular specialism with their own ideology. Within a particular profession there can be a series of loose amalgamations of practitioner segments that operate under a common institutional umbrella, the medical profession being a case in point. Power groups in professions will come together at an institutional level to present their own view of professionalism to the outside world. Over time, these powerful coalitions will change as individuals come and go. The profession, and its ideology, therefore evolves and changes over time to reflect changes in cabalistic power groupings.

To summarise, defining the boundaries of an occupation and profession is not easy. There are a number of different views on the nature of professions. However, the underlying themes are associated with power, status, qualifications, education, training and market inclusion or exclusion through the credentials offered by the professional qualifying association. The esteem within which the credentials are held is a factor of the public recognition attributed to the profession by society.

The next series of subsections explore the implications of the foregoing on the professional development of VM. The next subsection explores the knowledge base of VM.

The VM knowledge base – founded on theory?

Historically, VM, through the work of Miles, was founded on the technique of functional analysis. It can be argued strongly that this is the only distinguishing characteristic of VM from other management philosophies and approaches. Miles also proposed that VM should be conducted through the structure provided by the job-plan. Thus, Miles proposed both the technique and the process for the delivery of VM in a manufacturing context. Subsequently VM emerged into the construction arena and has given a new lease of life to it.

There is sufficient empirical evidence from this research project to indicate that the knowledge base of VM currently, in contrast to that proposed by Miles, is ill-defined or strays into that of other disciplines. These disciplines can be found within business schools or in construction-related departments in the higher education sector, such as architecture, surveying, construction management or project management. There have also been attempts recently to broaden the knowledge base of VM to include concepts from operational research, such as objectives hierarchies.

From this brief analysis it can be surmised that there is a theoretical underpinning to VM, as currently conceived, but it stems from and is borrowed from other discipline bases. In certain countries internationally, VM has been dominated by practitioner developments rather than through the interaction of practitioners and academics, the normal case within the traditional professions. Furthermore, in comparing countries internationally, VM has take root in academia in a different ways, and has embedded itself much more in some countries than in others. The diversity of the knowledge base, tools and techniques, is both a strength and weakness. The strength is the richness of tools, techniques and perspectives that can be drawn on to solve clients' value problems. The weakness is the lack of clarity in defining exactly what VM is, coupled with the potential for interdisciplinary and inter-professional 'turf warfare' over who 'owns' VM.

It can be concluded that the academic base of VM, as a discipline, is in its infancy. It has been dominated by practitioner developments, does not have a unique theoretical underpinning, but is founded on a unique technique and a process that is a good exemplar for problem solving. However, a foundation on a technique(s) and a process alone is insufficient to claim professional status. The important issue is the possession of a unique body of theory that guides practice. VM currently cannot claim this.

The individual practitioner knowledge and skill base

It has been stated earlier that the individual VM facilitator requires the following skills and knowledge base to undertake studies:

- experience
- competence in a variety of management skills related to human dynamics and the management of teams
- leadership qualities
- setting the workshop agenda.
- ability to manage the process of the workshop including:
  - dealing with any hidden agendas of participants
  - recognising individuals and their contribution
  - questioning and summarising
  - providing direction and a sense of common purpose
  - sensing interpersonal relationships within the team
  - sensing the climate of the workshop
  - synthesising and integrating information during the workshop
  - intervening in the workshop process when necessary
- communication skills, both verbal and for producing the VM report.

The preceding are generally high-level cognitive-problem solving and 'doing' skills that have their origins in the integration within the individual practitioner of education, experience, training and effective interpersonal behaviour.

The VM facilitator must also be trained in the VM process and its relevant tools and techniques. This links the individual practitioner back to the knowledge base of the service, which, as indicated, above is both diverse and rich.

It has also been alluded to earlier that knowledgeable, regular procuring clients have allegiances to particular facilitators rather than the employing consultancy. This indicates clearly that not everyone is capable of being a good VM facilitator and they are potentially a scarce resource in the market place.

There are a number of conclusions that can be reached at the individual practitioner level. First, for those clients that are convinced of the benefits of VM, good practitioners are in short supply and they therefore have an opportunity in the market place to establish an authority relationship between practitioner and client based on cruciality Second, and stemming from this, there is an inherent potential for good VM facilitators to create a mystique around the VM facilitation process since not everyone is capable of doing it and it requires high-level cognitive skills to work effectively in a workshop setting. The potentially high degree of mystique and cruciality provides the segment of good VM practitioners to establish a sound, potentially elitist, professional base for VM. It does, however, require an effective policing system to define the good VM practitioner from the not-so-good, not only for the continuing reputation of the service, the practitioners but also commissioning clients, either regular or irregular procurers of the service.

Professional associations, VM and professionalism

The study has identified the institutional and hence territorial framework within which VM develops in the UK as important for the delivery of the process. The institutional influences on and interest in VM in the UK is coming from a number of quarters, such as the Institute for Value Management, the construction professional institutions and academia. Equally, there have been other influences in terms of its deployment in the UK, such as an HM Treasury CUP guidance note, a European standard for VM, the 'Latham' Construction Industry Board Special Interest Group, the European Union SPRINT programme. These institutional frameworks at one and the same time expand, clarify, amplify, confuse and add to the rich development of VM in the UK. This subsection explores these issues in the context of the important role of professional associations for the future development of VM.

Construction industry professional institutions that have either laid claim to or are expressing an interest in VM are:

- Royal Institution of Chartered Surveyor (RICS). It has stipulated that it is one of its core services for the next millennium and has published numerous research papers on VM
- Chartered Institute of Building (CIOB). It has published research papers on VM
- Institution of Civil Engineers (ICE). It has set up a task force to produce a guidance note on VM
- Chartered Institute of Building Services Engineers (CIBSE). It has produced a guidance note on VM
- Association for Project Management (APM). It has a special interest group.

The Institute for Value Management, which encompasses construction as well as other industry sectors, has also indicated that management approaches such as business process reengineering, total quality management (TQM) and other related value oriented approaches, should also be part of the value manager's armoury. By broadening the potential toolkit of the value manager into these areas, the British Institute of Management (BIM) is also potentially placed within the orbit of those professional associations with an interest in VM. BIM also recruits members from the construction professions.

The RICS, CIOB, ICE, CIBSE and APM are all qualifying associations. They test competency and are actively involved in consolidating and developing the knowledge base associated with their occupational domains. As indicated earlier, qualifying associations also provide control over recruitment and training and they are also involved in the educational process. The RICS, CIOB, CIBSE and ICE, in particular, exercise occupational control over entry and the award of credentials at undergraduate degree level. As part of this process they will accredit or recognise undergraduate degree courses in tertiary educational establishments that meet their institutional criteria. Members of these professional associations who practice as value managers have already gone through a certification process for their own professions and it is not unusual for members of the APM to have multiple professional qualifications from different institutions due to the broad nature of project management. The fact that they are qualifying associations and have either committed themselves to VM as a core activity; have special interest groups in place or have undertaken and sponsored either research studies, publications or guidance notes indicates that there are segments within their membership with an inherent and substantive interest in the development of VM in their own professional arena.

In the UK the IVM is recognised by the EU Commission as the appropriate body dealing with VM matters in the UK. However, as indicated above, during the past four to five years a number of the professional institutions representing construction have also indicated a stake to the claim of the home of construction VM. This must have an influence on the debate on professionalism. The key question is that if a chartered surveyor, chartered engineer or architect were to profess a competence in VM and offer a service backed by the professional indemnity insurance required by these institutions, would a procurer of VM services feel comfortable about that person's professionalism? Alternatively, would that same procurer feel more comfortable with a professional value manager?

In applying the trait model of profession to VM the picture shown in Table 3 emerges.

The forgoing analysis indicates that the chartered institutes with an interest in VM are already well embedded with many of the traits of a profession and the associated trappings of professionalism to house VM as a specialist service from within their portfolio of services. However, to the best of the authors'

Table 3: Traits in VM

Trait	Comments
Self organisation through professional associations	In construction there are at least five associations with an expressed interest in VM. Of these some have well-established strategies and ideologies of professionalism in place. The IVM is the appropriate body recognised by the EU on VM related matters. It does not have a well-established system and ideology of professionalism in place. In comparison to the chartered institutes, the IVM has closer similarities to the APM, which also recruits across industrial sectors. Equally, a number of the chartered institutes have long-standing interests in project management.
Adherence to a code of conduct	The chartered institutes have established codes of conduct. This is also the case with the IVM.
Education and training, normally to degree level	The chartered institutes have long-established links, as qualifying associations, with the higher-educational sector. They also look to this sector, in addition to professional practice, to deliver training courses on a regular basis. VM has many affinities with project management, in that it is likely to be a graduate, post-professional activity given the requirement for experience. This suggests that VM requires a postgraduate delivery process. In the UK there are now many masters degrees in project management that have been established and running for, in some cases, decades. The same argument can be levelled at VM.
Complex occupation, with a knowledge base requiring specialist skills	The analysis earlier clearly indicates that VM does require the use of high-level cognitive skills, often to solve complex value-related problems. While it does require specialist skills they are not unique to VM, with the exception of functional analysis, and other occupations/professions can lay equal claim to the knowledge base once functional analysis has been mastered.
Skills based on theoretical knowledge	The answer is clearly "yes", but the theoretical knowledge base is not unique to VM. Again, other occupations/disciplines could lay claim to the knowledge base of VM.
A sense of vocation and belief in altruistic service	Most professions lay claim to altruistic service and this is probably best espoused as doing the best for the client in a business context. The research that formed the basis for this whole section identified empirically that a sense of vocation increases with age. It is probably a function of the opportunity cost of time and effort already sunk into a particular vocation compared to that of taking up another also having transferable skills. VM is probably one occupation that relies heavily on transferable skills.

knowledge, each of these associations does not have specifically training and certification programmes dedicated to VM, although this was recommended to the RICS in 1986.[16] They certify within their core disciplines but there may be only components of this process focusing on VM, if at all.

Furthermore, within the UK construction arena this debate brings into play the position of the National Certification Body for Value Management that administers the European Certification Scheme on behalf of the European Governing Board. Currently the secretariat is provided through the IVM, which has the advantage of representing VM in all industrial sectors. This links the IVM very clearly with the certification process of VM facilitators backed up by a European directive.

To add to the richness of the territorial debate over VM, SAVE International launched a global outreach programme at its 1997 conference with the specific aim of widening the influence and membership of SAVE internationally. As the original founding association for the value-oriented approach, coupled with the fact that there are a number of SAVE franchised associations internationally, this adds a further dimension and important player to the UK scene.

To conclude, the foregoing suggests no clear picture on the institutional framework for VM in the UK. What is clear, however, is the potential for territorial 'turf warfare' through inter-professional competition offset by the fact that the certification for VM is through a national certification body managed currently on behalf of the EU by the executive of the IVM.

Education and training

In the UK construction industry, VM education is now embedded within the tertiary educational fabric of this country, both at undergraduate and postgraduate levels. At higher-education level, universities, through their charters, are able to offer degrees in areas where they have expertise and there is a potential demand. Course development will often be linked into developments in professional practice or research, and this is especially the case within the domains of the traditional professions. In the construction area, there is normally a close relationship between the university sector and the professional qualifying associations. However, university charters permit academic institutions to act autonomously in the development and delivery of courses but in practice the close relationship between academia and the professional associations means there is a symbiotic relationship operating to the mutual benefit of both. The professional associations influence courses. Equally there is reciprocity in that academics are often members of committees within professional associations and this, in turn, influences their development. In terms of educational provision, at undergraduate level, the teaching of VM is likely to be at the level of awareness only. However, at postgraduate level, it is highly likely to be a combination of both education and training.

The charters of universities also provide for short continuing professional development courses to be delivered and many of the same principles cited in the previous paragraph may also apply. This may inevitably bring academics into direct competition with practitioners who deliver similar courses from within their own personal and organisational bases of expertise.

16. Kelly J.R. and Male S.P. *A Study of Value Management and Quantity Surveying Practice*. Surveyors Publications Ltd., 1988.

The symbiotic relationship between the university sector and the professional qualifying associations means that there are checks and balances in place in terms of the quality of delivery. This occurs at two levels. First, within the institution itself through its course vetting procedures and secondly, when the qualifying associations visit university departments to accredit courses, normally on a five year rolling basis. Shorter duration continuing professional development training courses will, however, normally stand outside of this accrediting process. However, certain qualifying associations in construction have approved centres for continuing professional development in the university sector and at postgraduate level they may well also accredit one or two year masters degrees.

The link between the theoretical base of a profession and its associated relationship with the education, research and training base within the university sector is important for the development of a profession. There is no doubt that VM in construction has been well established in the tertiary educational sectors of the UK and Australia for some time. In the USA it is to a much lesser extent. The traditional construction qualifying associations in the UK have been accrediting courses with VM embedded within the curriculum, certainly at undergraduate level, for a number of years. In the same vein, many have been sponsoring research studies in VM to support the continued development of the profession. This has also found its way into course development.

To conclude this subsection, VM has already fallen within the remit of many of the traditional qualifying associations in construction for a number of years. The higher education sector, seen as important for the continued development of a profession, is able through its statutes and charters to develop and deliver courses autonomously and has its own system of checks and balances in place to ensure quality of delivery. However, the symbiotic relationship existing between qualifying associations and the university sector provides enhanced checks and balances for the professional as well as the academic development of a discipline. From a UK perspective, it could be argued that the university sector, with a well established bedrock of VM teaching at undergraduate and postgraduate levels could proceeded alone in the educational and training areas. A precedent is available in Australia where the University of Canberra provides a certificated training course recognised by government departments. This operates outside of the professional association framework. For academia to pursue this strategy in the UK would, however, set aside the long-standing tradition of academics and qualifying associations working closely together to validate educational and training provision. Normal practice in higher education would suggest that any number of qualifying associations could accredit courses on VM, this would also include the VM National Certification Board or the SAVE International equivalent under their global outreach programme.

Certification revisited

Education is about imparting understanding and the ability to question, this may involve some training at undergraduate tertiary level and will almost certainly involve a greater integration between education and training at postgraduate tertiary level depending on the type of course. Training is concerned with transferring and developing skills, normally focused around the application of techniques. Training may not involve education at all but purely the teaching of technique. Neither equips a person for practice. Professional qualifying associations will normally accredit educational courses at undergraduate level (or possibly postgraduate level) but they will still require subsequently a period of practice before certification to practice is achieved. Certification for practice is held within the remit and delivered by the qualifying association.

Certification to practice VM, as argued in the report, requires the assessment of:
1.) the use high-level cognitive skills - given the nature of the skills required this could only be best achieved through some form of observation of the practitioner
2.) the application of the tools and techniques in the practice setting
3.) the extent of understanding of the knowledge base and its subsequent use in practice.

Over time the qualifying associations have developed tried and tested internal, self-regulated, quality-control procedures to achieve 1 to 3 (above), normally involving the interaction between practitioners, the qualifying association and academia. This is part of the professional development of an occupation and its notion of professionalism to safeguard client interests.

Currently, VM does not appear to have these mechanisms in place, regardless of which professional association has been mentioned. The market place is currently doing this, largely through the regular procuring client that have developed sophisticated approaches to vetting VM practitioners.

Summary and conclusions
There are clearly segments within existing construction professions, and expounded by their professional qualifying associations, that have a significant interest in VM. Equally, within the IVM it is likely that segments will emerge, if they have not already done so, arguing for specialist approaches to deal with, for example, construction, manufacturing or the service sectors. Taking the long-term view, it is unclear, as yet, which of the many institutes may form the natural home of VM or whether any one institute can claim a territorial right over it exclusively. However, what is certain is that institutional interest is moving forward, especially among some of those that have power and influence in the UK construction industry due to their historical longevity. Professional qualifications, professional responsibility and ethical behaviour are probably the keys to professionalism irrespective of the institutional home. A number of the construction associations espousing an interest in VM, and mentioned above, already have codes of conduct in place and a long-standing view of professionalism within their own professional domain. However, certification of VM currently rests with the National Certification Board operating under the auspices of the EU and via the IVM executive.

An argument can be set out that VM could be considered a profession in a country where the institute responsible for VM:
- becomes recognised by those practising VM, and those procuring VM services, as the accrediting body for courses of learning and for qualifications, i.e. the qualifying association
- is run quasi-democratically by the membership
- has a code of ethics
- has a clearly defined knowledge base and set of tools and techniques that are constantly in a state of ongoing development that differentiates it from other disciplines.

The argument against this is that VM facilitators do not offer advice nor provide a design, but they do provide a service and are liable for the quality of that service. If VM is a profession a certification scheme for practice is required. The system of training and certification proposed by the European governing board to be launched by the IVM in November 1998 is a scheme that has the essential ingredients of accreditation of education and certification. The Australian system is not strong because the accreditation is not organised by the IVMA but is instead organised by the Government of New South Wales via the University of Canberra. The system evolved by SAVE International meets all of the requirements but because the education element is kept in-house it could be seen to be too restrictive and inward looking. In contrast, the European system is more performance-orientated and therefore is more likely to be subjected to proper academic analysis.

The corollary is that VM is a profession then a standard is not required since certification is achieved through a self regulating qualifying association. A value manager operating to a particular national standard, however, is operating within an occupational framework, where service is ill-defined and the recipient needs some protection.

The next section discusses VM in thc public sector, where pressures on accountability and value-for-money are very strong.

5.6.4 VM in the Public Sector

There has been a substantial growth of VM in the public sector in the USA and in Australia, as indicated earlier. Conclusions with respect to the success criteria can be drawn as follows.

- A standard is useful in defining terms but is not critical to successful public sector implementation.
- A capital procurement procedure understood by procuring departments is more important than blanket legislation that requires the use of, *inter alia*, VM.
- In situations where the management of VM is not the responsibility of the procuring department the proper undertaking of VM by appropriately qualified VM facilitators is more likely.
- Appropriate qualification of those offering the VM service satisfies issues of public accountability.

In respect of the UK public sector, VM should become the responsibility of the organisation that sanctions the expenditure. This follows the successful model applied by New York City OMB.

New York City OMB is a part of the Treasury and is charged with the task of ensuring value-for-money on projects initiated by other departments within the city. A VM team within OMB receives details of all bids for capital works from the various city agencies. An assessment of the VM activity is made and contracts let to private-sector certified (SAVE International) value specialists. The project is then tracked and VM studies undertaken. The procurement and supply of VM services is by OMB. The VM studies are undertaken 'off-line' of the existing project using an independent team chosen and organised by the VM facilitator. It should be noted that the function of OMB is to track projects, initiate and pay fees for appropriate VM activity and to monitor implementation of the VM recommendations.

A similar situation exists within US Navy whereby a small team are notified of all capital works projects coming on stream. An assessment is made of the appropriateness of VM to the project and if appropriate then the number and timing of studies is notified to the initiating department. Arrangements are made and all fees paid by the VM Department. Again, similar to OMB, private sector VM practitioners are contracted on a framework agreement basis to undertake the planning and facilitation of individual studies undertaken off-line using an independent team. Overall, Navy spending budgets are not altered but often funds are redistributed through VM, usually resulting in more projects being brought on stream within any given financial year.

The implementation of VM within the UK public sector should be based upon a system that includes the following.

- A capital procurement procedure which incorporates the notification of projects to a VM department within the organisation which sanctions expenditure. This will enable the targeting of appropriate VM activity.
- The VM department should receive a budget for undertaking VM activity based on receiving a proportion of the total capital spend on construction. This removes the direct cost of VM from the public-sector agency initiating the project.
- The VM department should employ staff sufficiently skilled in VM activity to establish framework agreements with suitably qualified private sector VM practitioners. The department would have the responsibility of recognising and tracking the potential offered by VM on projects notified by procuring agencies.
- Agreements should be made with the procuring agency as to the timing of the VM exercises to be undertaken with the design team for the project, i.e. not using an independent team as OMB or the US Navy.
- Agencies should be assured that their budgets are not threatened through VM activity, i.e. that the intention is to ensure value-for-money not to cut costs and to potentially deliver a greater number of projects due to any savings identified.
- The VM department should track projects to completion and monitor the effectiveness and implementation of VM recommendations.

A policy such as that described above draws together the best of public-sector practice observed during this study.

6. Areas of Further Research

6.1 Benchmarking

The popular technique of benchmarking has found many uses within the management and production practices of some of the largest companies in the world. However, there has been no particular reference as to how this practice can be applied to a management or production practice whose output cannot be considered to be tangible. For example, VM executed at the concept stage of a project inevitably leads to design decisions being made to reflect the value system of the client. Unfortunately, since no prior design decisions have been taken it is impossible to determine how successful VM has been since the resultant design decisions cannot be measured back to any previous reference point. Conversely, implemented proposals resulting from the application of the VM process at detailed design will have the previous designs to measure and compare back to. This imparts the ability to measure tangibly the relative success of applying VM. It is proposed that investigating the use of 'soft metrics', such as customer satisfaction, for benchmarking purposes may allow the technique of benchmarking to expand into an area previously unexplored, with potential to reap more benefit.

6.2 Supply Chain Management

Previous work in VM has concentrated at one end of the supply chain and on the relationships between various sections of client organisations. Particularly, between the client and the design team and between the client, design team and contractor. This project has uncovered the application of VE techniques at the opposite end of the supply chain, between the suppliers and the

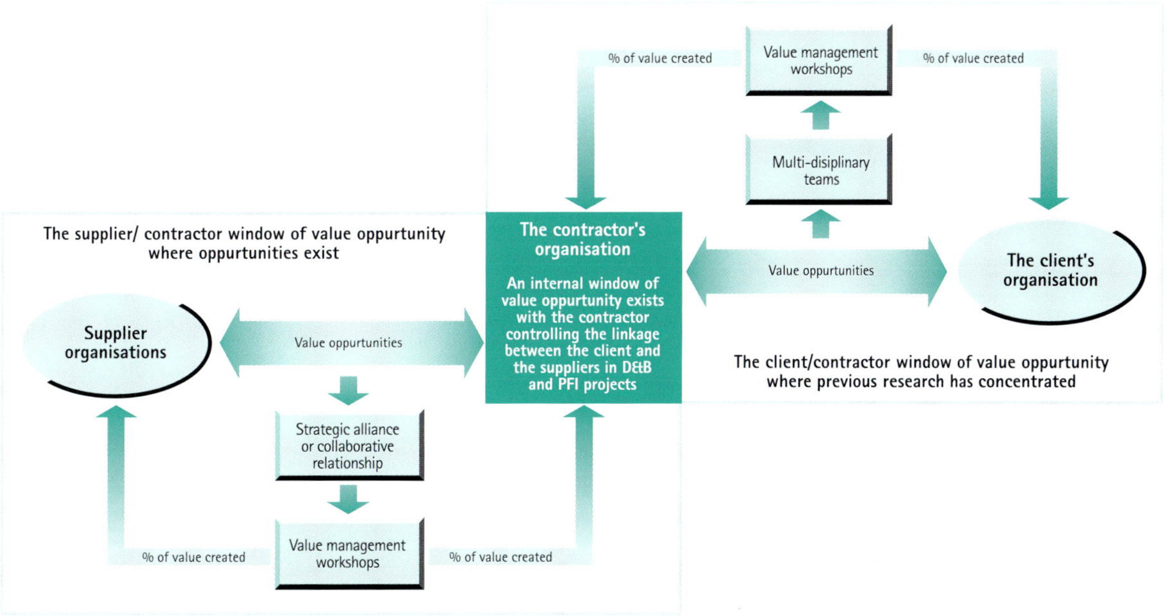

Figure 11 Value opportunities within the supply chain

manufacturing organisation, particularly in the automotive and electronic industries. This approach seemed to reflect an examination of the supply chain by the manufacturing organisations and the identification that the suppliers' costs account for approximately 70-80% of the total costs of the finished product. Subsequently, strategic alliances with a limited number of suppliers are formed to create a platform for the VE exercises to take place. These alliances with suppliers at the design stage, by making use of their specific expertise, result in large efficiency gains for both parties. The relationship is a 'win-win' situation where all parties benefit. Thus, further research into the application of this approach could prove fruitful to all participants in the construction process. This concept is detailed in Figure 11.

PART 3: An Analysis of VM Benchmarking in 1995: an Historical Background to the Project

7. A Review of Benchmarking

7.1 Benchmarking and Related Management Practices

The investigation of benchmarking was undertaken to determine the extent of knowledge and experience of it in the construction industry, and also to develop from the theoretical frameworks described in the literature a method of carrying out the process of benchmarking of VM. It should be noted that this is a historic view of benchmarking at the commencement of the project. The research progressed from this point.

Similarly, the management literature was examined to determine whether any of the techniques or characteristics of the VM process have been adapted from other techniques and developed under a new banner. However, despite the burgeoning number of new management techniques, VM remains a distinct discipline with its own identity. The importance that the EU SPRINT programme attached to VM lent weight to this assertion.

7.2 Background to Benchmarking

All the texts on benchmarking stress that while the techniques associated with the activity are not new, undertaking structured benchmarking is a more recent phenomenon. It is one of the most effective means to identify improvements that can make significant differences to the carrying-out of business processes. The process itself is twofold in that it involves the collection of process knowledge as well as the collection of performance data of the process being performed. In this way it is possible for benchmarking partners to identify superior practice through the use of metrics and to implement process changes to allow a comparable performance level to be attained.

It is generally considered that there are six basic steps to benchmarking:[17]

- deciding what to benchmark
- planning the benchmarking project
- understanding your own performance
- studying others
- learning from the data
- using the findings.

The research project followed this sequence of steps.

7.2.1 Common Types of Benchmarking

There are three common types of benchmarking.

- Competitive benchmarking - where information is sought from a hostile competitor. This is generally considered the most difficult type of benchmarking.
- Co-operative benchmarking - where knowledge is exchanged with the benchmarking target partner. In this type of benchmarking the target will usually give more information than they receive.
- Collaborative benchmarking - where there will be a full knowledge exchange between the parties.

17. Boxwell R.J. *Benchmarking for Competitive Advantage*. McGraw Hill, New York, 1994.

It was recognised that benchmarking VM was most likely to be a co-operative or collaborative exercise.

7.2.2 Success Indicators for Benchmarking

Camp[18] identifies ten success indicators which will ensure that a benchmarking exercise will run smoothly:

- an active commitment to benchmarking from management
- a clear and comprehensive understanding of how one's own work is conducted as a basis for comparison to industry best practices
- a willingness to change and adapt based on benchmark findings
- a realisation that competition is constantly changing and there is a need to 'shoot ahead of the duck'
- a willingness to share information with benchmarking partners
- a focus on benchmarking first on industry best practices and second, on performance metrics
- the concentration on leading companies in the industry or other functionally best operations that are recognised as leaders
- adherence to the benchmarking process
- an openness to new ideas and creativity and innovation in their application to existing processes
- the institutionalisation of benchmarking.

The whole exercise must be properly timetabled, with the identification of clear goals and deadlines. This is especially crucial when benchmarking 'customer satisfaction', seen an area of prime importance for VM. The preceding guiding principles for successful benchmarking were adopted as part of the research process.

7.2.3 A Benchmarking Code of Conduct

The benchmarking literature highlights the following code of conduct for benchmarking:

- all information gathering to be within legal limits
- the benchmarking partners to be prepared to fully share information and to keep the information received confidential if asked to do so
- the benchmarking partners to ensure that their respective expectations were fully understood
- the benchmarking partners to take an honest approach to the activity.

The VM benchmarking research project adhered to these basic codes of conduct throughout the study and provided potential partners with an outline of these prior to and during the benchmarking studies.

7.2.4 Benchmarking in the Construction Industry

UK managers have had little experience of the benchmarking technique in practice[19] but although this position is changing there is still a general lack of awareness of methods of structured benchmarking activities. Benchmarking, as a means of cultivating continual improvement, is emerging as one of the main tools of TQM and has a proven success in manufacturing and services industries in the USA and Japan. The growing body of management literature dedicated to benchmarking deals with good practice and describes a generic process model that can be used as a framework for carrying out benchmarking exercises.

18. Camp R.C. *Benchmarking: The Search for Industry Best Practice that lead to Superior Performance.* Quality Press, Milwaukee, 1989.

19. Lema N.M. and Price A.D.F. Benchmarking: Performance Improvement Towards Competitive Advantage. *Journal of Management in Engineering,* **11**(1), 28-37.

Recently the construction industry has made movements toward adapting and adopting many of the principles of TQM[20] but experience of benchmarking is limited. Of great importance in benchmarking is the need for improvements to be verifiable, which is possible through comparison of measurable performance metrics.

The literature review indicated that there is currently a lack of benchmarking standards for construction internationally, although industry and academia have recently turned their attention to their development. Lema[21] has proposed that a reason for construction lagging behind other industries in the use of benchmarking is that the concept and principles are more difficult to apply to essentially 'one-off', project based activities. In the USA the absence of any identifiable benchmarked data has stimulated researchers to compile initial data of construction activities for use by the construction industry.[22]

A common feature of the construction benchmarking activities described by Fisher[23] is that they are measurable in terms of cost and time, fulfilling a primary requirement of objective comparability. This type of data includes actual versus authorised project cost, actual versus target schedule, actual versus estimated labour, etc. These are all data that can be given classic statistical treatment and presented as averages, standard deviations, ratios, etc. At this stage of the research it did not seem realistic that meaningful external benchmarking of VM activities could be derived from this type of analysis due to the wide-ranging nature of the problems it addresses and the lack of regulation of its practice in the industry, i.e. VM is not sufficiently standardised to allow such a straightforward approach. The research team therefore investigated the development of its own approach adapted from benchmarking principles.

7.3 Benchmarking VM: the Measurement Issue

7.3.1 Introduction

The preceding sections have detailed the position of benchmarking theory at the commencement of the project. This section discusses the approach to the selection techniques appropriate to the benchmarking of VM. The section highlights some of the difficulties of applying the hard metrics, characteristic of traditional benchmarking practice, to the softer problems often encountered in VM exercises. The logical development of a benchmarking system appropriate for the research project is outlined.

7.3.2 The Characteristics of Benchmarking Applied to VM

The concept of benchmarking relies entirely on two inter-related and supporting pieces of data:

- a measure of performance for the process identified as the focus for the benchmarking study
- extensive knowledge by the partners to the benchmarking study of how that process is executed - it is usual for only two partners to be involved in a benchmarking exercise.

The concept is also based on the premise that carrying out the data collection and following it up with analysis will facilitate change leading to improved/superior performance within both organisations engaged in the process. In essence, if it can be determined through the collection of measurable process performance data that a process is executed better

20. Fisher D. *et al.* Benchmarking in the Construction Industry. *Journal of Management in Engineering*, **11**(1), 1995.

21. Lema N.M. and Price A.D.F., op cit.

22. Fisher D. *et al.*, op cit.

23. Fisher D. *et al.*, op cit.

elsewhere then there is scope for improvement for the organisation which does not perform that process to the level identified. However, this knowledge alone cannot facilitate change to improve and attain the identified performance level since there is no indication why a performance difference exists. Only by collecting supporting data on how the process is executed by the organisation performing it better can analysis support recommendations for change to improve performance.

The research project was initiated on the basis that the process knowledge of VM as understood by the research team could be enhanced through discussion with current VM practitioners and clients. It was always the objective that both the researchers and the practitioners would benefit from this interaction. It was anticipated that this style of investigation of VM would indicate a process performance metric.

An examination of the work of Fisher, described earlier, lead to the conclusion that it is not realistic to expect that meaningful external benchmarking of VM activities can be derived from 'hard' analysis. On the other hand the ultimate in 'soft' analysis, i.e. unstructured comparisons, could be criticised as being 'information grazing' or 'industrial tourism' from which few firm conclusions could be drawn.

The method chosen for the research involved stripping down VM into its critical success factors and then examining the hard techniques that lie within each factor. The hard techniques could be realised through an examination of case studies.

7.3.3 The Initial Proposed Methodology for Benchmarking VM

Reviewing the current application of benchmarking in industry, the collection of metrics during benchmarking exercises are based on hard tangible evidence of the performance of the processes involved. The processes benchmarked in these instances have outputs which are readily identified and measured in numerous ways to give an overall indication of performance, such as percentage of defects, cost, time taken, etc.

Applying the same principles to VM, it was thought possible to identify metrics which had the traditional characteristics that incorporated:

- identifiable techniques used in the process
- a quantified unit of measure
- a form which facilitated the construction of a meaningful database
- a capability of being compared across a range of VM workshops.

In this respect the following were considered.

- Number of ideas generated and workshop time, i.e. ideas/hour.
- Number of ideas generated and workshop cost to the client, i.e. ideas/unit cost.
- Number of ideas generated (costed or otherwise) and number of ideas submitted to the client or implemented - in effect an idea index measuring how successful creativity and evaluation techniques were during the workshop.
- Total cost of the workshop to the client.
- Proposed savings and proposed cost of implementation of ideas generated at a workshop, i.e. value for money. A quantified measure of workshop success. This is highly dependent on when in time the workshop is held - it

is generally difficult to quantify savings if the workshop is held early in the project life cycle but easier later in a project.

- The time taken for the workshop to come to a consensus view on, for example, client objectives, project objectives, etc., i.e. the speed of undertaking identified stages in the workshop.
- Efficiency of techniques employed during the workshop, e.g. FAST diagramming, brainstorming and methods of voting.
- Number of people in the workshop and background of participants.
- Time taken by the facilitator in information gathering and synthesis and workshop time.
- Time allocated for specific stages in the process as described by the workshop agenda, amended to reflect the reality of the workshop.
- The size and structure of the final workshop report.

In addition, there are a number of metrics which are softer in nature and can only be considered subjectively.

- Client satisfaction and piece of mind - satisfied that the proper design is being implemented.
- Value improvement - a better solution has been achieved with or without cost savings.
- Internal client, design and/or construction team building - the client's own team, the design and/or construction team coming away from the workshop working better as a team.

Having established possible hard metrics, the next section sets out the development of the research methodology subsequently.

7.3.4. Subsequent Stages in the Development of the Benchmarking Methodology Adopted

Benchmarking VM – the use of case studies

The following case studies are based upon an examination of completed VM reports of studies undertaken using the Kelly and Male methodology as VM consultancy projects. The aim of the case study investigation was to determine the extent to which VM studies could be the subject of benchmarking. The commonality of the authorship of the method, the facilitation of the VM workshops and the authorship of the workshop reports is a useful means of ensuring a background control and commonality of data for an initial trial, bearing in mind that:

- the workshop reports were contemporary with the workshop and were not written to facilitate benchmarking
- workshop reports are the prime source of data for benchmarking.

The case studies were chosen to represent VM studies at different stages in the project life cycle are as follows.

- Project 1: Pro-forma briefing document for a sheriff courthouse.
- Project 2: Design-build project at pre-bid stage for an industrial works upgrade.
- Project 3: Pharmaceutical clinical trials unit at outline sketch design stage.
- Project 4: Crown court at final sketch design.

The techniques used are summarised in Table 4.

Table 4 supports no specific conclusions. There are no relationships between ideas generated and the time spent at the workshop, time preparing pre-

Table 4: VM techniques applied in project case studies.

	Project 1: Sheriff Courthouse	Project 2: Works upgrade	Project 3: Clinical trials unit	Project 4: Crown court
Pre-workshop				
Post-occupancy evaluation of existing building	•			
Document analysis	•			•
Facility walk through	•	•		•
Interviews	•	•		•
Workshop				
Project context analysis	•	•	•	•
Function analysis		•	•	•
Function diagram	•			•
Time/cost/quality	•	•	•	•
User flows	•		•	•
Adjacency matrix	•			•
REDReSS	•		•	•
Brainstorming (ideas generated)	not recorded	392	205	90
Judgement	•	•	•	•
Development	•		•	•
Procurement choice	•		•	
Presentation	•			•
Workshop time: hours	30	12	24	40

workshop and ideas generated or tasks undertaken and ideas generated. There is some relationship between the number of techniques used and workshop time and possibly between time taken, the number of techniques used and overall workshop quality, although this can not be judged from the reports.

In judging the workshops, Project 4, the Crown court, could be said to be the closest to a textbook interpretation of a VM study aided perhaps by a previous study at outline sketch design stage. Project 3, the clinical trials unit, a study undertaken at outline sketch design stage, highlighted a number of logical concerns of a strategic nature, mainly related to time and cost certainty coupled with a risk averse client. These issues should strictly have been addressed earlier in the project life cycle. Project 2, commissioned by the design/build contractor with the co-operation of the client, was concerned with generating as many ideas as possible and getting client feedback. The other tendering contractor was also given the opportunity of undertaking a VM exercise. Project 1, the Sheriff courthouse, was an exercise at producing a generic brief and as such had no specific project applicability. A part of the information stage was undertaken as a POE of an existing Sheriff courthouse.

Stage 1 – outcomes following the result of the case study experiment

The literature on benchmarking raised a number of issues of which the dominant are seen to be:

- identifying the critical success factors and determining which of these to benchmark
- establishing quantifiable performance metrics (defined as measurable outcomes that indicate a degree of success in achieving some NM objective.

Initially, from a review of the action-research consultancies conducted by Kelly and Male and cited above, the research team identified the critical success factors as:

- structuring the project information
- the structured approach through the 'job-plan'

- the skill of the facilitator
- the team structure, skills and working relationship established during the workshop process.

The attitude of stakeholders to the project

However, the dilemma was "what to measure?" The case studies supported the conclusion that the analysis of a complete VM study is unlikely to yield any benefit. This was a disappointment as the majority of VM practitioners keep detailed reports of their VM workshops that would have provided a rich source of data. It was also recognised that all construction projects are unique and the case studies demonstrate that fact. This is not the case for benchmarking application in the manufacturing sector since the product remains constant and forms the basis of most comparisons.

It was not felt that VM is a unique process in this respect and work by others such as Ziari and Hutton have identified that there exist problems for benchmarking in the areas of innovation and new product development. This led the research to conclude that the application of benchmarking to these areas needs to address the measurement issue and identify suitable 'soft' metrics for use in these circumstances since there exists no 'hard' measure of performance.

Stage 2 - the case vignette approach

Following the deficiencies highlighted in the analysis of case studies it was decided to compile a number of case vignettes of typical projects. These case vignettes were drawn up to reflect projects of different complexity, size, client type and value. Interviews were held with practitioners to determine the VM approach in each case. After a number of interviews it was found that the approach taken to each project type represented by a case vignette was the same. Therefore, complexity, size, client type and value did not significantly affect the approach to VM taken by the practitioner. The only active variable was the stage the project had reached, i.e. the approach taken would be different for a project a briefing stage to that taken for a project at sketch design stage. The length of the workshop also was more correlated to stage of the project than for example project value.

Stage 3 - making explicit the Kelly and Male methodology

It was decided following Stage 2 to undertake a detailed breakdown of the Kelly and Male methodology with specific reference to techniques used at the different project stages. This is described in detail earlier in the report. Structured interviews were then held with practitioners where the methodology was presented and discussions took place on the perceived benefits of various approaches, tools and techniques.

This was the process of benchmarking adopted.

Stage 4 - validating the results

It was realised that the extensive database of techniques derived from the interviews had no performance measure and the case study analysis described above had eliminated the majority of performance measures. This left customer satisfaction, a 'soft' metric as the remaining metric to indicate the performance of the process. It was reasoned that the success of VM could be judged by how well the client perceived the VM process or how well the project performed after the application of VM. The research team concluded that an indication of how the client perceived the success of the VM workshop would be a good indication of the performance of VM application. The collection of data on the success of the project after the application of VM could be added to give an overall indication of the performance of VM undertaken in practice.

Having identified a metric it remained for the research team to collect this data. However, the only means of distributing objective questionnaires to clients was via practitioners due to commercial confidentiality. Therefore questionnaires relating to customer satisfaction had to negotiate a complex international route. The final outcome was that insufficient data was collected for analysis.

In addition, the project had to adopt a slightly different approach to benchmarking from the outset since unlike the traditional application of benchmarking between two partners, it was intended by the research to employ the use of numerous partners. It also differed in the respect that because the Kelly and Male methodology was subject to incremental improvement following each benchmarking interview the research team were not in effect one of the partners themselves. The research team was in effect a third-party facilitator of the benchmarking process applying it across numerous partners.

Validation of the study was undertaken through seminars comprised of experts from all sectors, including practitioners and clients. The framework document was, after some revision, accepted as the 1998 benchmark of VM.